Why Business Ethics Matters

Why Business Ethics Matters

Answers from a New Game Theory Model

Wayne Nordness Eastman

WHY BUSINESS ETHICS MATTERS
Copyright © Wayne Nordness Eastman 2015

All rights reserved. No reproduction, copy or transmission of this publication may be made without written permission. No portion of this publication may be reproduced, copied or transmitted save with written permission. In accordance with the provisions of the Copyright, Designs and Patents Act 1988, or under the terms of any licence permitting limited copying issued by the Copyright Licensing Agency, Saffron House, 6-10 Kirby Street, London EC1N 8TS.

Any person who does any unauthorized act in relation to this publication may be liable to criminal prosecution and civil claims for damages.

First published 2015 by
PALGRAVE MACMILLAN

The author has asserted his right to be identified as the author of this work in accordance with the Copyright, Designs and Patents Act 1988.

Palgrave Macmillan in the UK is an imprint of Macmillan Publishers Limited, registered in England, company number 785998, of Houndmills, Basingstoke, Hampshire, RG21 6XS.

Palgrave Macmillan in the US is a division of Nature America, Inc., One New York Plaza, Suite 4500, New York, NY 10004-1562.

Palgrave Macmillan is the global academic imprint of the above companies and has companies and representatives throughout the world.

Hardback ISBN: 978–1–137–43043–4
E-PUB ISBN: 978–1–137–43045–8
E-PDF ISBN: 978–1–137–43044–1
DOI: 10.1057/978–1–137–43044–1

Distribution in the UK, Europe and the rest of the world is by Palgrave Macmillan®, a division of Macmillan Publishers Limited, registered in England, company number 785998, of Houndmills, Basingstoke, Hampshire RG21 6XS.

Library of Congress Cataloging-in-Publication Data

Eastman, Wayne Nordness.
   Why business ethics matters : answers from a new game theory model / Wayne Nordness Eastman.
       pages   cm
   Includes bibliographical references.
   ISBN 978-1-137-43043-4 (alk. paper)
   1. Business ethics. 2. Game theory. I. Title.
   HF5387.E27 2015
   174'.4—dc23   2015013924

A catalogue record for the book is available from the British Library.

This book is dedicated to Hal Pond Eastman, Jr. (1930–2012)

Contents

List of Figures	ix
Preface	xi
Acknowledgments	xv
Overview of the Book	xix
Introduction: The Four Temperaments and the Four Games	1
Part I Humors and Games	**19**
1 We're Better Than We Think	21
2 The Harmony Games	39
3 Opening the Door to the Sanguine	61
4 Bringing Telos Back	93
Part II Business Ethics	**117**
5 Critical Business Ethics	119
6 Why Business Ethics Matters	141
Conclusion	161
Appendices	163
Notes	175
References	185
Index	197

List of Figures

	The Four Temperaments	xix
I.1	The Four Temperaments and the Four Games	16
1.1	Ignorance Is Bliss	37
2.1	The Weakest Link	59
3.1	Alternative Stories	91
4.1	Telos	114
5.1	The Blame Game	139
6.1	A History of Ethics	159

Preface

In Part One of this book, I advance a game-theoretic version of the classical four temperaments perspective on human nature. In Part Two, I offer an understanding of business ethics as a phlegmatic, pragmatic, and practical way of solving social games that is more productive than, but not morally superior to, other ethics animated by more emotionally intense temperaments.

Writing this book has involved a very long journey, in which I have experienced my own versions of the classical repertory of Sanguine, Melancholy, Choleric, and Phlegmatic feelings. Over the years on that journey, I have been inspired by the scholarship of four teachers I have been lucky enough to have known, and who have served as intellectual lodestars. Through them, I have learned, succeeded, failed, and tried again.

Professor Thomas Schelling, who many years later won a well-deserved Nobel Prize, was the first of my four guides at Harvard College in 1973. I felt tremendous enjoyment in my freshman seminar with him that fall.[1] I was fascinated and excited by the idea that you could use game theory to understand the world. I loved the 2×2 matrices he introduced us to, and the sometimes logical, sometimes psychological exercises he had us do and discuss.

In particular, I was deeply impressed by the disturbing logic of the Prisoner's Dilemma that Professor Schelling described to us. How could rational egoists escape the trap of following a "dominant strategy" that made both players better off no matter what the other did, yet left them both with a poorer outcome than they could have if they'd only been able to cooperate?

The man with the unfashionable bristly crew cut and glasses who was teaching us was one of the leading strategists of the Cold War era, when thousands of American and Russian missiles were poised to strike the other nation's cities and people at a moment's notice. By 1973—thanks in part, I believe, to Professor Schelling's work—detente was in the air, and nuclear war had become a less omnipresent and frightening prospect than it had been in the early 1960s, when my elementary school classmates and I had hidden under our desks in simulated fallout drills at the time of the Cuban missile crisis.

After enjoyment came shame. Like many academically inclined undergraduates in the 1970s who in another era might have gone on for their doctorates, I felt dubious about academic life and went to law school. PhDs were driving taxicabs, or so rumor had it, because the professorial jobs were all taken—and wasn't the real world the place to be in any case, not the ivory tower? But my dreamy, theorizing side remained strong, and in law school, I encountered the second person who transformed my thinking about games and the world.

Professor Schelling had made the eighteen-year-old me a deep-dyed believer in game theory. Duncan Kennedy—a charismatic, long-haired Harvard Law School professor who was a star in the then-new Critical Legal Studies movement—helped make the somewhat older me a skeptic about standard game theory, and about my earlier enthusiasm.

My loss of faith did not come from personal preaching by Duncan, for I never took a class with him, but from his articles, and, perhaps, through some instant mind-meld, from a time I saw him give a talk.[2] Post-Duncan, I was still preoccupied with the Prisoner's Dilemma. But now it was a skeptical, debunking fascination. As I toiled away as a litigator at a Wall Street law firm in the go-go 1980s, I consoled myself with the prospect of collecting my bonus, quitting my job, and writing a genre-busting philosophical novel that would include a critical dissection of game theory.

Hoping to make a break from law practice to teaching, I sent out letters to various schools inquiring about job possibilities and describing my novel. No job interviews resulted, but I did get a short anonymous note, postmarked from Michigan, saying that I should take a look at Robert Frank's *Passions within Reason*.[3]

I duly read the book. I was extremely impressed at Bob's account of how moral emotions could serve strategic functions—for instance, of how blushing could be a reliable signal of a character that was embarrassed by a lie and thus likely to be a trustworthy trading partner. Stimulated by *Passions*, I read other academic and popular works by Bob that used the Dilemma in a variety of imaginative ways to argue for public polices—for example, for work safety regulations as a good way to control a race to the bottom based on workers caring about their financial position relative to other workers, and hence valuing safety too little.

In the 1990s, my split "Duncan" and "Bob" halves were both productive. With the help of a business law colleague of my father's, I'd been lucky enough to get an adjunct teaching job at my father's school that eventually led to a tenure-track job. Faced with publish-or-perish pressure, I buckled down to write two kinds of articles. One kind drew on my "Bob side" to make a game theory–based case against the rat race. For example, I surveyed my MBA students on their work hour preferences and analyzed the results to support the claim that managers in general, and women managers with children in particular, were trapped in a Dilemma that led them to work longer hours than they preferred.[4] The second kind drew on my "Duncan side" to make a case that logical models usually associated with one political position—the Prisoner's Dilemma and liberalism, supply–demand curves and free-market conservatism—could be flipped to tell the opposite side's story.[5]

My split halves had worked fine for writing articles. But I could not make them pull together in 1999 after I got tenure and had a year-long sabbatical, or over the next ten years or so that I struggled futilely with successive versions of what was supposed to be a book on political ideology. There was a division within me. Was I tearing the heart out of ideology, and the fever dreams of believers of all stripes? Or was I supporting ideology, and advancing my own "neither right nor left" ideology? The result was a hypertrophied righteousness module. I was grumpy about my intellectual guides, and righteously angry about my own and everyone's self-righteousness, partly because I couldn't acknowledge and accept the ashamed, fearful, sad side of myself.

In the last four years or so, I believe I have gotten some way to the balance that so long eluded me.

First, reality brought me closer to accepting sadness as a part of my life. The Saturday of Memorial Day weekend in 2010, my sister in Massachusetts called to tell me that my father had had a seizure while driving with my mother, had been diagnosed by the doctors at Newton-Wellesley Hospital with a brain tumor, and would be operated on at Mass General Hospital on Monday. His tumor, we learned after his operation, was an invariably fatal glioblastoma, the same type that killed Teddy Kennedy.

My father died in 2012. One Saturday morning in April he was walking around a pond, his optimistic spirit if not his mind intact. A day and a half later he was gone. Sadness remains.

A second reason for possibly moving closer to balance involves a shift in my teaching and research focus over time from business law to business ethics. For me, law, like politics, powerfully stimulates the point–counterpoint, righteousness-first part of myself. Ethics, not so much. Happiness surfaces more easily; competitive fervor is less powerful.

Another reason I think I have come closer to balance relates to a fourth intellectual mentor, social psychologist Jonathan Haidt. I've gotten to know Jon in his new job as a business ethicist at New York University, where he runs a seminar with Bob that I attend. Jon's work criticizing (and appreciating) righteousness among political believers and all the rest of us helped me to let go of my ambition to write a politics book, and to turn my book-writing focus toward business ethics.[6]

At long last, forty years after my fall afternoons in Cambridge with Professor Schelling, I am closer, I hope, to the spirit of calm and appreciation of all four of my intellectual mentors that I need to write a book that draws from them. What I have to say combines their modern approaches to games, to social science, and to criticism with a very old understanding of ethics as balance that is found in the classical West and also, in somewhat different versions, in other parts of the world. After reading this book, you will be able to draw on a new way of understanding ethics in general, and business ethics in particular, in terms of temperaments and games. That understanding may, I hope, be of assistance to you in attaining your own version of balance at home and at work.

Acknowledgments

To turn this book into reality, I have needed to learn from many people. I needed the guiding stars I wrote about in the Preface, Thomas Schelling, Duncan Kennedy, Robert Frank, and Jonathan Haidt. The idea of viewing the world in terms of games that I learned from Professor Schelling; the idea of flipping that I learned from Duncan; the idea of useful moral emotions that I learned from Bob; the idea of joining tradition with evolutionary psychology that I learned from Jon: Without these ideas, and the embodiment of them in people I looked up to and aspired to follow, I could not have written this book.

I have also needed to learn from my colleagues. Michael Santoro, with whom I have enjoyed conversations and collaborations for twenty years, was indispensable. By providing numerous close, searching, and helpful comments, and by assuring me that I already had the book written at a time when I felt very far from completion, he helped make his faith in me, and the book itself, real.

With Michael, I am very fortunate to be part of an excellent group of business ethics faculty at Rutgers Business School, which now includes Danielle Warren, Chris Young, Ann Buchholtz, and Michael Barnett. I am grateful to all of them, and to Danielle and Chris in particular for their many hours devoted to critiquing my work, and helping to make it better.

The pivot in my intellectual interests from law and politics to ethics and business that made this book possible could not have taken place without the leadership of my now-retired colleague Ed Hartman in building an ethics faculty at Rutgers. My midlife awakening to Aristotle and virtue ethics owes everything to Ed.

For her consistent support of my research and for providing an outstanding exemplar of calm, Sanguine leadership as department chair of Supply Chain Management and Marketing Science, and since the beginning of this year as the new Dean of Rutgers Business School, I am very thankful to Lei Lei. The interdisciplinary department that she built and that I have been happy to help administer as vice chair has been a great home for developing the ideas that turned into this book. All of my departmental colleagues have been helpful; particular thanks go to Arash Azadegan and Kevin Kolben, who have kept me on my toes in numerous conversations devoted to this book and other topics, and to my companions in department administration, Rosa Oppenheim, Jacqueline Perkel-Joseph, Corinne Schiavo, Dottie Torres, Shen Yeniyurt, and Yao Zhao. I appreciate the helpful feedback I received on the book, and/or on projects that led to it, from colleagues in the department and other departments, as well as from colleagues in other schools, students, family, and friends: Ali Asani, Nicole Bryan, Bruce Buchanan, Chao Chen, Laura Chinchilla, Deirdre Collier, Jennifer Crohn, Emilio De Lia, David Dobrzykowski, Carolyn Eastman, Carroll Eastman, Rebecca Eastman, Stephanie Eckerd, Bob Frank, Georgiana Hart, Leon Fraser, Pierre Gagnier, Michelle Gittelman, Jon Haidt, Darcy Hall, Jonathan Hall-Eastman, Caroline Hall-Eastman, Mary-Ella Holst, Nien-he Hsieh, Duncan Kennedy, Rose Kiwanuka, Terri Kurtzberg, Don Klock, Farrokh Langdana, Eric Larson, Rudi Leuschner, Joe Markert, Ben Melamed, Yaw Mensah, Sean Pidgeon, Anne Quarshie, Guy Quinlan, Rusty Reeves, Dale Rogers, Gene Spiegle, Erich Toncre, Can Uslay, and Tom York.

For over two decades now, Rutgers Business School has been a congenial home for me. My father first started teaching at the school in 1964, and I grew up doing exercises that he gave to his organizational behavior students. Fifty years later, I am proud to be continuing his legacy. This is my book, but it is also his. His spirit as an optimist, organization man, and rebel—the spirit that I absorbed sitting at his knee with my sisters, listening to him strum the guitar and sing his business school song, "The Administrative Point of View"—is the spirit I have tried to give voice to here.

I am thankful to the academic business ethics community, embodied in the Society for Business Ethics and its journal, *Business Ethics Quarterly*. I want to single out two editors of *BEQ*, George Brenkert and Gary Weaver, who provided careful and critical feedback on articles that are part of the background to this book. I am also very thankful to Ed Hartman, Bob Frank, Jon Haidt, Bruce Buchanan, Eric Schoenberg, and the other organizers and supporters of the Paduano business ethics seminar at NYU, which for years has been an intellectual second home for me, and where I presented an early version of the ideas in this book.

Part of the book is rooted in a review article and other projects related to critical business ethics that I have been working on over the past few years. My key collaborator in these projects over the past two years has been Nicole Bryan. I am very grateful to her, both for her contributions to the critical business ethics enterprise and for her peripatetic energy in connecting with people, which has served as an inspiration for me. I also want to acknowledge other scholars with whom I have worked on critical business ethics papers and presentations: Chris Young, Laura Chinchilla, Anne Quarshie, Sasha Poucki, Rose Kiwanuka, Yao Zhao, Kevin Lyons, Jason Stansbury, and Ed Wray-Bliss.

Another inspiration for this book lies in civic, political, and religious activity I have been engaged in over the past twenty years in my hometown South Orange, its sister community Maplewood, and its urban neighbor Orange. There are many people I have collaborated with, competed with, or both in those endeavors from whom I have learned, and to whom I am appreciative. Particular thanks go to Madhu Pai, Jeffrey Bennett, Andrea Marino, Mark Gleason, Jennifer Payne-Parrish, Andrew Lee, Fred Profeta, Don DeMarco, Barbara Heisler, Andrew Lee, Darcy Hall, Aisha Hauser, Yielbonzie Johnson, and Darrell Berger.

Producing a book is a team enterprise; I am very thankful to Leila Campoli of Palgrave Macmillan for taking a chance on me, to Mary Child for valuable editing and advice, and to Laura Chinchilla for the same.

I am also thankful to the hardware and software, to the many places, and to the people at those places, that were partners in getting this book written. Without the Phlegmatic Harmony I enjoyed with Dell, Typepad, Open Office, Microsoft Office, and a wide variety of workplaces, ranging from the Elephant House cafe across the park from my daughter Caroline's flat in Edinburgh, to my Rutgers offices in Newark and Piscataway, to coffee shops and libraries in South Orange, Maplewood, Plainfield, Newark, Summit, Westfield, Chatham, Orange, Madison, Montclair, New Brunswick, Little Falls, Paterson, Manhattan, Brooklyn, Barcelona, Singapore, and other locations, the book might well have never seen the light of day.

I am deeply thankful to my wife Darcy Hall, the person who more than anyone else over the last thirty years has supported and sustained me. I am also thankful to my children Jonathan and Caroline, who grew up hearing about different versions of this project. In recent years as adults, they have helped me grow up with them, and have given me much appreciated intellectual, emotional, and technical support on this final version.

Finally, I am thankful to all the writers itemized in the references, and to many more who are not, from whom I have learned. As one who has lived my life in considerable part through books, I feel extremely grateful to them, and to my mother for encouraging me as a young child to read, and to walk alone down the street to the library. I am very happy to be contributing at last to the great table at which I have supped for so many years.

Overview of the Book

The Temperaments	Active/Yang !	Reactive/Yin . . .
Positive :)	Sanguine :) !	Phlegmatic :) . . .
Negative :(	Choleric :(!	Melancholy :(. . .

The Four Temperaments

The overarching idea of the book, illustrated in the figure above, is that we have an ethical nature that accords with the classical view of ourselves as divided into four temperamental quadrants. We have a Sanguine, happy quadrant; a Phlegmatic, practical quadrant; a Choleric, angry quadrant; and a Melancholy, sad quadrant. The basic reason for our four-part nature, I will suggest, is to equip us with the emotions and intuitions that allow us to solve four major kinds of social games that correspond to the four temperaments. Our cheerful, optimistic side enables us to do well in Sanguine, or Harmony, games; our pragmatic, calm side helps us in Phlegmatic games; our anxious, ashamed side helps us get along as well as may be in Melancholy games; and our righteous, punishing side helps us cope with Choleric, or Disharmony, games.

I have been inspired in writing this book by my students, and by the experience of being a teacher. As I worked to finish the book, I taught business ethics to executives in Singapore and to MBA students and undergraduates in New Jersey, and also taught ethics to second and third grade children in a religious education program. In what follows, I draw on all these classes, and on my years of teaching experience. Much as this book is highly abstract and theoretical in some respects, it is also intended to be highly practical, and useful to teachers and students. With that in mind, at the end of the introduction and each chapter I will offer suggestions on how the material in the section can be taught, whether to others or to oneself, and taken to heart.

Introduction: The Four Temperaments and the Four Games

In this book, I suggest that if we bring together the modern system of game theory with the classical system of the temperaments, or passions, we can make progress in understanding our ethical nature, which is not possible with either system alone. In realizing how happiness, anger, calm, and shame all help us solve social games, we can attain a better grasp of the logic of human social interactions and of all kinds of social interactions, including our sometimes frustrating, sometimes satisfying interactions with nonhuman actors, such as software programs and organizations. Together with other people, we can draw on our intuitions, our emotions, and our reason to do a better job in creating Harmony[1] with people, with nature, and with our material and abstract creations, in different moods—tranquil, compliant, competitive, and, especially, happy.

By combining game theory with the temperaments, we can also make progress in understanding business ethics as both a state of mind and an historical phenomenon. Business ethics, in the view that will be advanced here, is a distinctive way of solving social games that relies especially on the Phlegmatic,[2] practical, and pragmatic side of our nature. It is now historically ascendant, and has helped bring about our highly productive material and cultural order—but it is not, I contend, morally superior to other temperamentally based ethics that have been ascendant in earlier eras of human history.

The strategy of the book is inspired in part by the modernist interpretations of the classical Four Temperaments by George Balanchine, Paul Hindemith, and Tanaquil LeClercq in the 1946 ballet of that name. Balanchine's choreography and Hindemith's music—flowing and quiet in the opening Phlegmatic variation, jittery and aggressive in the closing Choleric variation, danced by LeClercq—were combined with simple black and white costumes of leotards and T-shirts, at the time a radical innovation in classical ballet. The ballet was better, more informed critics than

I have agreed,[3] for not simply mimicking a classical worldview and classical balletic technique, but for fusing it with a stripped-down, abstract modernism. A similar point applies here. I hope, in this book, to marry good, true, and beautiful elements in the classical tradition of the temperaments with the same elements in another version of twentieth-century modernism: game theory.

At nearly the same time as Balanchine's ballet—in the deep ethical shadows cast by World War II, Hiroshima, and the Holocaust—John von Neumann, Oskar Morgenstern, John Forbes Nash, Thomas Schelling, and other intellectual pioneers developed a new, abstract rhetoric of matrices and mathematics.[4] They used this new rhetoric to analyze the logic of social interactions of all kinds between all types of people and entities: prisoners, spouses, teenage daredevils, business competitors, the United States, and the Soviet Union. Their new rhetoric combined abstruse, esoteric language with powerful, highly disturbing hypothetical scenarios: the Prisoner's Dilemma, the Battle of the Sexes, Chicken, and the Stag Hunt. All of these stories—especially the uber-story of the new rhetoric, the Prisoner's Dilemma—embodied a troubling message about the tension between logic and virtue: To be perfectly informed and logical—to clearly understand one's interests, and to act on that understanding in a perfectly rational, calculating fashion in a game with another person who was equally perfect—was to fail in achieving what you and the other player desired together. Your individual interest and your collective interest were at war. Rationality and ethics, far from being conjoined, were locked in struggle.

More than sixty years have passed since the doom-haunted 1940s that gave birth to game theory. From its origins as a field centered on calculating human actors, game theory, in its rising, evolutionary form, has broadened its focus.[5] It now offers itself to us as a way to understand the logic of all kinds of interactions—games—between entities of all kinds, whether or not they are calculating, conscious, or alive. This book is animated by a hope that the mood of our time has shifted sufficiently from the appropriately sad, self-reproachful one that ruled when game theory was born, and was reflected in the work of the pioneers of the field, to a more open-ended, balanced spirit. In our time, it has become possible, I hope, to advance a new, non-mathematical, humanistic, optimistic, temperament-based interpretation of game theory, one that is respectful of the calculating, relentlessly logical, and self-critical spirits that ought to rule in their place, but also of other, freer, more cheerful spirits that have their own proper domains. Advancing such an interpretation is the major intellectual project of this book.

The Four Games

[I]t is happy for you that you possess the talent of flattering with delicacy. May I ask whether these pleasing attentions proceed from the impulse of the moment, or are the result of previous study?
—Mr. Bennet to Mr. Collins, from Jane Austen,
Pride and Prejudice (1813)

What is a game? Some of us, such as the enigmatic philosopher Ludwig Wittgenstein, view the concept of a game in very broad terms indeed. Language is a game, and so, too, is reason. I embrace such a broad view of games, but will not argue for it here. For our purposes, what is important is that a game is not just an activity like chess, soccer, or StarCraft, with more or less clear rules, a winner, and a loser. If that were so, critics would be right in claiming that game theory implicitly or overtly advances a competitive and egoistic view of the world, given the centrality of beating the other player, or team, for board game athletes like Bobby Fischer, ball field athletes like Lionel Messi, and video game athletes like the StarCraft champion Scarlett.[6] But alongside the obvious, and correct as far as it goes, interpretation of a game as a competitive activity, an alternative interpretation of a game as a cooperative, social activity is also possible, and is in fact widespread.

Games may involve ritualized practices like table manners or the differing protocols for strangers, acquaintances, and friends addressing one another. In other words, games may involve social activities, practices, customs, or conventions in which social concerns such as mutual advancement, self-respect, respect for the other, sympathy, empathy, and avoiding shame or embarrassment are at stake. In the Jane Austen passage above, the adroit Mr. Bennet is playing a social game with the maladroit Mr. Collins—indeed it is a social metagame, with the two of them discussing how social games are played. In non-zero-sum, social games like the one that Messrs. Bennet and Collins are playing, we have important starting points for an ethically complex game theory that is aligned with the multiplicity of our moral emotions.

One reason for optimism about the possibilities for a new form of game theory arising now is the rise of new forms of games over the last few decades. When von Neumann, Schelling, and others did their pioneering work, zero-sum, competitive games like poker occupied the attention of people who liked to play games and think about them. Today millions of people spend many hours playing games like Sim City[7] that are noncompetitive, and sometimes cooperative. Many others play games like MineCraft,[8] in which non-competitive and cooperative strategies coexist with competitive ones, in a way that has substantial parallels to the complexity of motives and behavior in the real world. Much as "destroy the foe" games like StarCraft and many others continue to uphold the competitive conception of a game, the twenty-first-century world of gaming we now live in is a much more diverse, open-ended one than the mid-twentieth-century world in which classical game theory was born. In the growth of diversity in gaming over the last few decades, we have, I believe, a significant foundation for the growth of diversity in game theory.

A Prequel of the Four Games

The central idea of this introductory chapter, as of the book as a whole, is that human nature can be understood in terms of four emotion-laden, intuition-driven temperaments that have evolved to solve four major kinds of social games. Two of our temperaments, traditionally called the Sanguine and the Phlegmatic, are positive. In the classical typology that is adopted here, we have an active, energetic, sociable, happy part of ourselves—the Sanguine—that helps us do well in games in which the players have shared feelings and interests. Such pleasant, happy games, which here will be called Harmony Games, have not been studied enough in classical game theory, I believe; they are the central subject of the first two chapters.

Second, we have a reactive, peaceful, calm, calculating, practical part of ourselves—the Phlegmatic. This part corresponds well to solving what here will be labeled Imperfect Harmony, or coordination, Games, represented by the classical game theory stories of the Stag Hunt and the Battle of the Sexes, which will be addressed in Chapter Three. In these games, the players' feelings and interests are fairly well, but imperfectly, aligned. A calm, forbearing disposition helps the players avoid the pitfalls that more intense emotions can create in these games.

In the Four Temperaments model of our nature, we also have two negative sides that correspond to two difficult kinds of games. We have an active, driving, competitive, angry, punishing, self-punishing part of ourselves—classically, the Choleric temperament. This part helps us deal with what are called Disharmony Games here, such as the Prisoner's Dilemma,[9] in which there is a serious conflict between the players as well as a shared interest in cooperating. The players, whether altruistic or egoistic, have a clear course of action. Unfortunately, it leads them to do worse than they would if only they could throw logic to the wind. In these disturbing games, anger at an uncooperative player, including self-punishing, guilty anger at oneself, can help resolve the situation as well as it may be resolved.

Finally, we have a reactive, negative, anxious, ashamed, obedient, withdrawing part of ourselves—classically, the Melancholy temperament. This part helps us to deal with what I call Partial Disharmony, or hierarchical, games, such as the classical stories of Chicken and Hawk-Dove. In these games, yielding to a dominant first mover makes sense. A compliant, passive, sad part of ourselves is helpful in resolving such games as well as they may be resolved.

In the second part of the introduction, we turn in more detail to the four temperaments that correspond to the four games. Before doing so, a few observations on the four games are in order. First, conventional game-theoretic treatments of social games rely on mathematical reasoning and often, though not always, assume that the players have reliable information about payoffs. Such standard treatments are typically unsystematic, in that they involve presentations of a grab bag of stories with a mixed set of messages about the failure or success of egoistic calculation in particular contexts, without any underlying typology of games. By contrast, the critical, or flipping, approach to games advanced in this book assumes that we are almost always unclear about others' payoffs and, for that matter, our own. This approach classifies social games systematically, relies on ordinary non-mathematical language, and tells stories that focus on the failures as well as the strengths of pro-social motivations like altruism, deference, and shame.

Second, although I have referred in a few places to the classic game-theoretic stories, I am very much committed to not telling, and not dwelling on, these stories at this point. The reason is that traditional stories—though not the underlying logic of game theory itself—assume egoistic or competitive motivations on the part of their protagonists—the prisoners, daredevils, and so on. Much as such motivations are a very important part of our nature, so too are socially oriented motivations, such as respect, sympathy, loyalty, shame, and guilt. If the alternative version of game theory advanced in this book is to live, it needs to have its own stories with its own characters. Moreover, these stories need to live in our minds, not simply as abstract alternatives to the standard stories of egoism and competition but rather as full-blooded descriptions of life at work and elsewhere.

A Prequel of Alternative Game-Theoretic Stories

A major theme of this book is that we can, and should, flip standard game-theoretic stories that assume egoistic and competitive motivations to create alternative stories that assume pro-social motivations of various kinds. In these flipped stories, problems for the players arise, just as in the standard stories. But, I suggest, we can reason out how to rectify these issues for pro-social players in a more positive, accepting spirit than we are likely to do with the problems of egoistic and competitive players. I give some sample stories here, all of which will be developed in detail in Chapter Three.

The first story, Deference, is a flipped version of the most famous traditional story, the Prisoner's Dilemma. In Deference, both the players want to help the other player. For both of them, Deferring by playing a second-fiddle, a helper's, role is better, no matter what the other does. The rub is that they, along with their firm and society, would both be better off if only they could both Assert instead of Deferring.

It is not necessarily any easier to solve Deference than it is to solve the Dilemma—lack of leadership may be no easier to cure than selfishness. There is a very big difference, though, between the feelings that attend the two stories. In Deference, you are unlikely to feel anger at the helpful other for failing to Assert, while you are very likely to feel anger at selfish behavior by the other in the Dilemma. You can think about how to solve Deference in a sympathetic spirit, rather than in the punishing spirit that often makes sense in the Dilemma.

The second story, I Get It!, starts with a failed Harmony, or Imperfect Harmony, game in which you are mad because you feel the other player has been a fool by not understanding what is in his own interest, as well as yours and everyone else's. In the I Get It! flip, you view the other player not as an idiot but as someone trying to help you, who, understandably, does not know best how to do it.

The third story, Gratitude, starts with a successful Harmony game played at a bakery, in which you calmly and contentedly view yourself and the baker as guided by the pursuit of self-interest that works for both, in the tradition of Adam Smith's invisible hand. In the flipped version, you view yourself and the baker as guided by the desire to please and other pro-social feelings. You experience different, more joyous, emotions, which may trigger in you new ideas about how to run your business and your life.

The fourth alternative story, Managerialism, starts with an unsuccessful game in which a CEO and a governing board you serve on are bogged down in busywork and in which you view both the board and management as driven by the desire to dominate the other. In the flipped story, you take a very different approach. You explain the managerial swamp you and the CEO are in as being the result of your shared commitment to hard work and agreeableness. In that flip, you gain insight into possible ways you can change to share leadership instead of busywork, without the change involving the recriminations that accompany standard stories of self-aggrandizing, slacking managers and board members.

The Symmetry of Altruism and Egoism

The difference between the conventional treatment of social games and the alternative approach advanced in this book involves a broad, conceptual dimension, as well as the more specific dimensions that have just been discussed in the flipped stories. The most famous standard game-theoretic story, the Prisoner's Dilemma, is typically understood as an illustration of how we often need to rise above our self-interest to achieve a good outcome. In the critical version of game theory here, embodied in the Deference story, the conventional interpretation, valid as it is on its own terms, misses a more interesting point about the Disharmonious logic that rules both the egoistic world of the Dilemma and the pro-social world of Deference.

In Disharmony, the altruistic reason of the deferrers falters, just as much as the egoistic reason of the prisoners does.[10] In fact, there is an exact logical parallel between the strengths and weaknesses of altruism and those of egoism in solving games. For every game in which altruism fails, such as Deference, there is another, exactly parallel one, in which it succeeds, such as the Prisoner's Dilemma. And so too with egoism. It must be so. If player A and B are both playing as pure altruists, they are in effect switching identities, with A becoming B and B becoming A. In doing so, they face exactly the same conundrums faced by the egoistic A and B.

Starting with that critical realization of the symmetry of altruism and egoism, we can, I suggest, move to affirmative strategies that take pro-social intuitions and feelings in ourselves and other people seriously, and that allow us to rise above the widespread errors of regarding such pro-social feelings as immune to the pitfalls of egoistic and competitive ones, and as only marginally relevant in the real world. Pro-social intuitions and feelings are very important, and are prone to their own major failings in practice. Armed with self-awareness, we can proceed to apply reason to remedy the failings of pro-sociality, just as we can for the failings of egoism and competitiveness.

There are some significant limits to the optimistic self-improvement and social improvement project that is suggested in this book. The emotional, intuitive humors that drive us in both pro-social and other directions are not purely, or even mostly, under our control. Indeed, as Robert Frank has argued, convincingly to my mind, what makes our passions work to solve games is that they have a life of their own, beyond our conscious direction.[11] But neither are we simply pawns of our mutable humors. Informed by a better apprehension of our complex, divided selves, we can make progress in figuring out ways to lead better lives at work and elsewhere, I maintain.

The Four Temperaments

> Yet is not the power which injustice exercises of such a nature that wherever she takes up her abode, whether in a city, in an army, in a family, or in any other body, that body is, to begin with, rendered incapable of united action by reason of sedition and distraction; and does it not become its own enemy and at variance with all that opposes it, and with the just? Is not this the case?

Yes, certainly.

And is not injustice equally fatal when existing in a single person; in the first place rendering him incapable of action because he is not at unity with himself, and in the second place making him an enemy to himself and the just? Is not that true, Thrasymachus?

—Plato, *The Republic* (B. Jowett, trans.)

Here then I find myself absolutely and necessarily determined to live, and talk, and act like other people in the common affairs of life. But notwithstanding that my natural propensity, and the course of my animal spirits and passions reduce me to this indolent belief in the general maxims of the world, I still feel such remains of my former disposition, that I am ready to throw all my books and papers into the fire, and resolve never more to renounce the pleasures of life for the sake of reasoning and philosophy. For those are my sentiments in that splenetic humour, which governs me at present.

—David Hume, *A Treatise of Human Nature* (1739)

In the first quote above, Socrates is responding to the blustery Thrasymachus. Thrasymachus has claimed that justice is simply the interest of the stronger, cloaked with a false mantle of righteousness. Socrates's response boils down to injustice being a losing strategy: Compared to their ethical counterparts, the unjust city, group, and individual are all torn apart by internal division and lack the capacity for effective action. Or, to translate Socrates's claim, which is also a fundamental claim of this book, into the language of evolutionary psychology and evolutionary game theory that will be used here: *Ethics is a successful, stable strategy at the individual as well as the group level.*[12]

At the time when Plato wrote the *Republic*, the belief that there are multiple humors, or temperaments, within a physically, psychologically, and ethically healthy self, reigned in Greece. It also prevailed, in related, though somewhat different, forms, in Indian Ayurvedic practice, in Chinese medicine and philosophy, and in belief systems elsewhere in the ancient world. That union of medicine, psychology, and ethics lasted for millennia, until relatively recent times.[13] The fundamental idea that all the main temperamental qualities, including the negative ones of anger and sadness as well as the positive ones of serenity and sociability, are necessary parts of a healthy, ethically sound human nature is the backdrop of Hume's quote. In it he describes his own quicksilver transitions—from melancholy speculation to Sanguine sociability with his companions to indolent, Phlegmatic acceptance to splenetic ire.

With the rise of modern medicine, and more broadly of mechanistic modern science, the idea of a balance among shifting temperaments—Sanguine, Melancholy, Phlegmatic, and Choleric—dominant as recently as in Hume's eighteenth-century Edinburgh, has faded.[14] More broadly, the belief that not only people but also other animals are made up of the four humors, and that inorganic matter is made up of the four related elements of air, water, fire, and earth, has become a pseudo-scientific rather than a scientific position. Still more broadly, the idea of a great chain of being that unites humans with nature in a purposive whole, in which each being has an ethical end, has become a matter of faith rather than one of reason.[15] Given these changes, the case Socrates makes for ethics lacks the scientific foundation for modern readers that it had for the classical Greeks, and for others in the premodern world.

I have no desire in this book to rehabilitate the classical four elements of physics, or premodern Western medicine and its doctrine of a physiological need for balance among the humors of blood, black bile, yellow bile, and phlegm. But at the heart of this book is a claim that classical psychology and ethics were on to a very important truth about us: For people to do well and do good in everyday life, they indeed need to have multiple, very different moral emotions, or temperaments, and to have these emotions in some kind of overall balance. Moreover, the moral emotions that help people succeed in the major types of social games are basically those of the classical four temperaments.

As discussed, the happy, active, positive Sanguine temperament is particularly useful in helping us create, and take advantage of, pleasant Harmony games, in which the players' feelings, interests, and intuitions are aligned. The positive, reactive, cool Phlegmatic temperament is helpful in solving Imperfect Harmony games, in which one good turn brings about another. The negative, reactive, ashamed, Melancholy temperament helps in resolving difficult Partial Disharmony games without undue conflict. Finally, the negative, active, punishing, self-punishing Choleric temperament helps enforce decent behavior in highly difficult Disharmony games like the Prisoner's Dilemma.

In the following chapters, I suggest, relying on game theory and psychology, that being an ethical person is advantageous not only at the group level but also at the individual level. I want to persuade you that being an ethical person helps individuals succeed not only in long-term relationships with other people or groups—"repeat games"—but also in "I'll probably never deal with this person or business after this" situations—"one-shot games."

The reason, I suggest, is fundamentally the one Socrates offers in the *Republic*. The ethical person, unlike the unethical person, is in a state of psychic balance or equilibrium that makes him or her effective in social games.

I believe that the truth of ethics as a healthy human nature, with different, shifting humors making up that nature, has been lost more in the West than elsewhere in the world, and more among highly educated Westerners than among others. For all its very great value, the relentless spirit of logical inquiry that one finds in Socrates and in his philosophical and scientific successors in the West has, I believe, created difficulties for highly educated Westerners. Their—our—deep preoccupation with being right, with overriding our emotions and intuitions ("System 1," to employ Daniel Kahneman's useful term)[16] in favor of reasoning ("System 2"), and with superseding the superstitious, pseudo-scientific elements of our past, makes it hard to see the very important elements of psychological and moral truth in what our ancestors believed. Reason, I suggest in this book, should lead us back to classical times as well as forward.

* * *

Id–Ego–Superego Compared to the Four Temperaments

In order to properly understand the Four Temperaments view of our nature advanced in this book, it is useful to view it in comparison with two other prominent conceptualizations of the human psyche. In one major classical and modern view of human nature, we are divided between a raw, emotional, selfish, hedonistic sub-self—the id or "it," to use Freud's term—on one extreme, and a guilty, ashamed, fearful, anxious, rageful, punishing sub-self—the supergo—on the other extreme.[17] In the middle is the healthy, rational sub-self, the ego. We are the charioteers of Plato's *Phaedo*, holding the reins of reason on the two great horses that pull us forward, but also threaten to tear us apart. In a second prominent conception of our nature, advanced by Hume and recently articulated by social psychologist and business ethicist Jonathan Haidt, reason is dethroned. Instead of being the driver, reason is a mere rider on the elephant of emotion and intuition. It is an instrument of the passions, not their master.[18]

First, let's compare the Four Temperaments view to the Id–Ego–Superego view. While in the Id–Ego–Superego view, the oversocialized superego and the undersocialized id are separate, opposite entities, in the view advanced here all the temperaments contain within themselves both a pro-social and an antisocial component. We have an active, negative, driving, angry, punishing, self-punishing Choleric sub-self that can be either highly socialized, like the angry superego, or highly antisocial, like the rageful id. We also have a reactive, negative, ashamed, anxious, fearful, obedient, withdrawing, depressed Melancholy sub-self, which also may be either highly socialized, like the compliant superego, or antisocial, like the infantile id.

In the Id–Ego–Superego view, the negative sides of human nature are a bug, rather than a feature. That leads to a clear normative recommendation, epitomized in Freudian psychology: We should struggle against the extremes of the undersocialized id and the oversocialized superego, on behalf of the healthy ego in-between. In the Four Temperaments view, by contrast, the negative is positive. The Choleric and Melancholy elements of human selfhood, negative as they are, are very useful in solving social games, and are to be cherished just as much, or nearly so, as the positive elements. They are brilliant, beautiful, good, and true, as well as profoundly disturbing.

Now, the positive side of ourselves. In the Four Temperaments view of our nature, reason is found in all of our quadrants. But the simultaneously calculating and social form of reason that is central to business has a particular connection, in the interpretation advanced in this book, to the Phlegmatic, pragmatic, practical, calm, stolid, soldiering-on part of ourselves. This reactive, positive, comparatively tranquil, instrumentally reasoning part of our selfhood is not the charioteer of our unrulier sub-selves. But it is neither simply a rider on them nor a servant to them. It is its own, coequal component of our complicated nature.

Finally, we have the star of the temperaments: the active, positive sub-self, the Sanguine. Here is where we have everyday happiness, momentary ecstasy, and joy that can never be permanent—for the Sanguine side of ourselves is eternally changing and evanescent—but that can be a greater component of our lives, as well as a lesser one. This sub-self, like all of the temperamental quadrants, has both a highly socialized side and a highly individualized, self-concerned side. Our Sanguine sub-self revels in being with others, in feeling their joys and their pains, and also revels in the self, and in its sometimes antisocial passions.

In the Sanguine, and its slightly-above-the-rest position, lies the greatest difference between the Four Temperaments view and the Id–Ego–Superego view. The Id–Ego–Superego perspective on our nature implies pessimism about two of the three parts of ourselves. By contrast, the Four Temperaments view advanced here embodies a fundamental optimism about all the major parts of ourselves, combined with a hope that the Sanguine part might in various ways move more to the fore in our individual lives, in our relations with other people, in our organizations, and in our societies than it has hitherto.

* * *

She whose complexion was so even made
That which of her ingredients should invade
The other three, no fear no art should guess;
So far were all removed from more or less.
—John Donne, *The First Anniversary: An Anatomy of the World* (1611)

> *She, she, is gone; she is gone; when thou knowest this;*
> *What fragmentary rubbidge this world is;*
> *Thou knowest, and that it is not worth a thought;*
> *He honors it too much that thinks it nought.*
> — John Donne, *The First Anniversary: An Anatomy of the World* (1611)

The Ideals of Balance and Purity, Personified

Within the classical Four Temperaments view and the modern version of it that will be advanced here, there is an overarching ideal of balance among the temperaments with a modest edge for the Sanguine over the others. The ideal of balance is the background to Donne's celebration of the equable mixing of the elements in the recently deceased Elizabeth Drury, in the first passage above. The ideal of balance coexists with an ideal of purity, of each temperament or combination of temperaments, having its place and its time, in which it reigns, with other temperaments excluded. In Donne's second passage, it is the purity of the bereaved's negativity that makes his feelings resonate—or Harmonize, as I will be putting it—with us as his readers. If the recollector's Melancholy, with its tincture of Choler, were dampened by ambivalence, whether of a Phlegmatic nature—"In the eyes of reason, what does my sadness matter?"—or of a Sanguine nature—"All is not lost, for there is salvation"—we would have impurity rather than purity. It would result in a less well-crafted and less emotionally powerful passage, one that would lack the resonance that Donne's actual language has with a Melancholy, Choler-tinged part of ourselves.

Proponents of the Id–Ego–Superego view of the self, with its clear affirmation of ego over the other elements, may see the coexistence of the ideals of purity and balance in the Four Temperaments concept as a sign of logical inconsistency. We must rank and decide! And indeed, sometimes we must. But in the Four Temperaments view, the imperative mood is only a part of us, and only a part of ethics, not the ruling spirit. Instead of a ruling "thou shalt"—"thou shalt obey the moral law; thou shalt maximize utility; thou shalt be a good person"—the Four Temperaments perspective as interpreted here offers us an alternative way to understand ethics. In this understanding, the imperative mood is only one of the humors. The imperative is a Melancholy, "we must do this or be ashamed" call that is very fine in its time and place, but not as the perpetual lord of the self.

It is important not to confuse the Four Temperaments approach advanced here with personality psychology,[19] which for all its considerable value is not the subject of this book. Though the version of ethics advanced in this book is a comparatively relaxed one, it has some bite. For a person to recognize himself as having a nature unduly titled toward Choler is to attain a self-critical insight. By contrast, a person who scores 100 percent on, say, intuition as opposed to sensing on the Myers–Briggs scale[20] is unlikely to regard himself self-critically. Allowing for the major point that this is a book about ethics, not about personality, I would suggest that the simultaneous elevation of balance and purity in the Four Temperaments view is best understood not in terms of argumentation or abstraction, but of personification.

Whether in an individual, a group, a firm, or a culture, a healthy balance of temperaments is the ideal—but the temperamental purity, clarity, and, perhaps, crazy intensity in an individual, a group, a firm, or a culture is also an ideal, and is what lends that entity vividness and force. We can think of the four temperamental quadrants as being embodied in philosophers, such as Aristotle, Bentham, Nietzsche, and Pascal, in great women, and goddesses and gods, like the martial Choleric Kali, the Melancholy Mary, the practical Phlegmatic Jane Austen, and the Sanguine Zeus, as well as in demigods of popular culture, like musicians and movie stars.

Particularly relevant for our purposes in this book are business and political leaders. The driven, angry, competitive, Choleric temperament? Much as his management style embraced other elements, Jack Welch of GE commanded ire as one key part of his arsenal.[21] How about the optimistic, upbeat, social, self-concerned, changeable, joyous, Sanguine spirit? Two twentieth-century presidents of different parties, Ronald Reagan and Franklin Roosevelt, stand out as exemplars. The sad, ashamed, lonely, Melancholy spirit? The leader who was possibly the greatest one ever produced in America, Abraham Lincoln, stands out on this dimension. And finally, what about the calm, practical, stolid, instrumental, pragmatic, Phlegmatic spirit? This last and least vivid of the temperaments, the one that corresponds closely, I believe, to business and to business ethics works well with the great calculating investor Warren Buffett.[22] We may, and we should, dispute the simplification inherent in these four-part classifications—but in understanding them, we give ourselves the basis to move forward in our inquiry.

What Lies Ahead

In Part One, "Humors and Games," I present a perspective on ethics that views human beings as intuitive creators of shared-interest games—Harmony Games. The central idea of the first two chapters is that we need to understand Harmony Games and their benevolent, happy logic, just as we need to understand the less benevolent logical conundrums of the difficult games focused on by the founding figures of classical game theory, which are treated in Chapters 3 and 4.

In Chapter One, "We're Better Than We Think," I suggest that human nature can be understood as a conspiracy to Harmonize: We excel at solving social games with other humans intuitively through interpreting our ambiguous feelings and complex social realities in Harmony terms. I suggest that the social convention of Harmony has played a central role in our success as a species, relating my claims to evolutionary psychology.

In Chapter Two, "The Harmony Games," I connect Harmony to the classical four temperaments, drawing on literary and popular fiction as well as other sources. I explain how Harmony Games, though centrally infused with a Sanguine, happy spirit, can also be played in other temperamental modes. I defend Harmony against a claim that it is ethically neutral or trivial, and suggest that we consciously devote ourselves to the creation of everyday Harmony Games.

In Chapter Three, "Opening the Door to the Sanguine," I describe how conventional game-theoretic stories, such as the Prisoner's Dilemma, that assume egoistic or competitive motivations can be reversed, or flipped, into new stories, such as Deference, that assume altruistic and other pro-social motivations. Compared to conventional game-theoretic stories, these alternative stories can help us to appreciate our situations and to reason in an equable, happy spirit, I claim.

I extend the scope of the inquiry in Chapter Four, "Bringing Telos Back," the most speculative part of the book. Here, I relate the Four Temperaments approach to game theory to nonhuman game players, including yeasts and businesses. I suggest that animals, objects, and organizations solve social games, and can reasonably be viewed as sharing an ethical nature with humans. Evolutionary game theory can help, I suggest, to give us the intellectual resources to create our own version of the purposive cosmos that ancients like Aristotle expounded.

In Part Two, "Business Ethics," I move first to a close-up perspective on business ethics in the classroom and in practice, and then back out to a wide-angle historical perspective on the field. In Chapter Five, "Critical Business Ethics," I suggest that we can usefully apply the Four Temperaments approach to business ethics situations. I give classroom exercises, provide a Harmony-based challenge to the classic business ethics portrayal of an inherently unethical corporate culture that needs to be defied, acquiesced in, or exited, and offer thoughts about potential research directions in business ethics and management.

In Chapter Six, "Why Business Ethics Matters," I propose that business ethics be understood as an historical phenomenon, as well as a state of consciousness. I suggest that business ethics is a currently ascendant, pragmatic, Phlegmatic way of solving social games that is more productive, but not more morally elevated, than different social ethics that have been ascendant during earlier periods of human history. I suggest that we can understand ourselves as being divided psychically now, as in the past, into:

- a *Choleric* quadrant that is especially useful in solving Disharmony Games, and that was dominant in the egalitarian, anti-oppression,

warrior ethics ascendant in the long hunter-gatherer era of human prehistory;
- an ashamed, *Melancholy* quadrant that is especially useful in solving Partial Disharmony Games, and that was dominant in the hierarchical, priestly ethics ascendant in the agricultural era;
- a *Phlegmatic*, pragmatic, practical, productive quadrant that is especially useful in solving Imperfect Harmony Games, and that is dominant in the currently ascendant system of business ethics, as it is defined here; and
- a *Sanguine*, sociable, happy-go-lucky quadrant that corresponds to Harmony Games and that may possibly have its day in a future society of abundance and long lifespans.

In this introduction and all of the chapters of the book, I conclude with a summary figure and a summary of the section's main idea, along with suggested exercises. Much as I hope each word of this book will be enjoyed, drawn upon, reflected upon, and argued with at leisure by some of its readers, I also hope that it will be worth sampling by modern-era readers with little time, or a preference for the visual and the intuitive over slabs of text. For such readers, and for old-fashioned word-by-word ones as well, I hope the figures, summaries, and exercises will be useful resources.

An Aspiration

As one who feels an affinity for eclectic, socially oriented, character-oriented classical virtue ethicists like Aristotle, Plato, Confucius, and Lao Zi,[23] a hope of mine is that this book can help to address an issue that has always concerned me about their work. The narrow, principle-based, "thou shalt" modern moral lenses of Kant's deontological, duty-based ethics and Bentham's utilitarianism[24] have been accompanied over the past few centuries by beautiful, intricate, "System 2" towers of reason like the Napoleonic Code and the Arrow–Debreu model of general equilibrium. By contrast, the broad, eclectic, optimistic, socially oriented "System 1" spirit of virtue ethics has a tendency to shade into commonsense bromides, and to be accompanied by less impressive edifices of reason than those erected by votaries of the narrower, principle-based modern approaches.[25]

I believe that the freer, more open-ended, more commonsensical spirit of virtue ethics should prevail on the whole in my field of business ethics over the narrower, more compulsive, more technical spirits of utilitarianism and deontology. At the same time, though, I also believe it is incumbent upon those of us who support virtue ethics approaches, broadly defined, to engage in projects of advancing our own versions of empirical and analytical reason.[26]

The relational, interpersonal, Four Temperaments version of game theory advanced in this book is one such project. My hope is to provide a more rigorous, though not unduly rigorous, contemporary way to articulate the ancient, enduring, approach to ethics that upholds balance among multiple goods—positive feeling and negative feeling, the reactive yin and the active yang,[27] Sanguine beauty, Phlegmatic truth, Choleric justice, and Melancholy compassion—over moral absolutes (Figure I.1).

The Temperaments	Active/Yang !	Reactive/Yin . . .
Positive :)	**The Sanguine Game: Harmony**	**The Phlegmatic Game: Imperfect Harmony**
Negative :(	**The Choleric Game: Disharmony**	**The Melancholy Game: Partial Disharmony**

There is an affinity between the four temperaments and the four games. The active, positive Harmony Game shares those qualities with the Sanguine humor; the active, negative Disharmony Game shares those qualities with the Choleric humor; the reactive, positive Imperfect Harmony Game shares those qualities with the Phlegmatic humor; and the reactive, negative Partial Disharmony Game shares those qualities with the Melancholy humor.

Figure I.1   The Four Temperaments and the Four Games

Summary

Human beings can be understood as divided into four roughly equal parts, which we can call temperaments or humors. Each of these parts has an affinity with four major types of social games that are important in business and in other domains. We have an active positive quadrant, which, following traditional usage, we can call the Sanguine; an active negative quadrant, the Choleric; a reactive positive quadrant, the Phlegmatic, and a reactive negative quadrant, the Melancholy.

The active, positive, Sanguine part of ourselves corresponds to Harmony Games, in which the players have shared feelings and interests. The active, negative, Choleric part of us corresponds to Disharmony Games, such as the famous Prisoner's Dilemma, in which the interests of the individual and the group are opposed. The reactive, positive, Phlegmatic side of us corresponds to Imperfect Harmony Games, in which the players' feelings and interests are well but imperfectly aligned. Finally, our reactive, negative, Melancholy side corresponds to Partial Disharmony Games, in which the players' feelings and interests are substantially but not completely misaligned.

(continued)

> *Summary* (continued)
> A view of human nature that emphasized different temperaments, or humors, within us, and the need for balance among them, was dominant for thousands of years in different forms around the world, and was linked to physics, medicine, and ethics. The view advanced here discards pseudo-science in the classical view, but endorses its fundamentals. A useful way to understand the Four Temperaments view of human nature advanced here, and to remember the four quadrants and their associated games, is in terms of personification. We can see real and mythical individuals with vivid, powerful personalities, or worldviews, as embodying a temperament.

> *Exercises*
>
> As an aid to absorbing the material in this introduction, and in the other parts of the book, I believe it can be valuable to engage in Internet-aided group or individual exercises, which can be done with the leadership of a teacher in a physical or virtual classroom, or by an individual student or reader anywhere. My suggestion is that such learning can usefully be divided into four categories: (1) exercises related to people, feelings, and relationships that draw from academic, literary, or "high" arts and letters; (2) exercises related to people, feelings, and relationships that draw from popular applied sources; (3) exercises related to concepts, things, and animals that draw from business, science, and technology; and (4) exercises related to concepts, things, and animals that draw from popular applied sources. In what follows, I give examples from each category; many of them are drawn from my experience teaching business ethics. Some sources for the exercises are in the endnotes and the references; others are available online through Google or another search engine.
>
> 1. (a) Watch and discuss, or reflect on, a part or all of Balanchine's "The Four Temperaments"; watch and discuss, or reflect on, a part or all of a documentary on Tanaquil LeClercq, who started dancing the final Choleric part of the ballet as a fifteen year-old; (b) With a focus on the Four Temperaments, read aloud, and then discuss, or reflect on, a part or all of Donne's "Anatomy of the World"; (c) With a focus on the relationship between System 1 and System 2, read and discuss, or reflect on, a part or all of Sylvia Nasar's biography of John Nash, *A Beautiful Mind*; (d) Read and discuss, or reflect on, the passage from Hume's *A Treatise of Human Nature* in which he discusses his moods and connects them to his philosophizing.
> 2. (a) Watch and discuss, or reflect on, videos of business leaders that highlight different temperaments; videos on Jack Welch, Al Dunlap, and Steve Jobs are good for showing the Choleric side at work, as is a video showing an exchange
>
> (*continued*)

> *Exercises* (continued)
>
> between Milton Friedman and a student on cost–benefit analysis and the Ford Pinto; (b) Search for, take, and discuss, or reflect on, online quizzes that provide scores on the relative levels and the balance among one's Sanguine, Phlegmatic, Melancholy, and Choleric elements; (c) Relate Four Temperaments quizzes to other personality or character quizzes, especially those based on Hans Eysenck's translation of the temperaments into introversion-extroversion and neuroticism-stability dimensions, David Keirsey's translation of the temperaments into behavioral roles, and Isabel Briggs's Jung-inspired focus on cognitive styles.
> 3. (a) Read and discuss, or reflect on, the part of von Neumann and Morgenstern's *Theory of Games* that discusses the game-theoretic problem posed in Arthur Conan Doyle's story "The Final Problem"; (b) Watch and discuss, or reflect on, all or part of Thomas Schelling's 2005 Nobel speech; (c) Read and discuss, or reflect upon, all or part of Daniel Kahneman's *Thinking Fast and Slow*.
> 4. (a) Do the quiz in the endnotes on the four kinds of social games and the emotional responses they provoke; (b) Search for, discuss, and/or reflect on online video games and the balance of different temperamental elements in them; (c) Watch and discuss, or reflect on, Tim Harford's video on Thomas Schelling.

PART I

Humors and Games

CHAPTER 1

We're Better Than We Think

This opening chapter tells a basically optimistic story about human nature that centers on the concept of Harmony Games in which people help other people and help themselves at the same time. In the account to be given here, we intuitively identify with other human beings. We are species-ists, and our species-ism is basically a good thing. Our ability to treat other people as "we," with other animals and nature as "them," has played a central role in our becoming the planet's dominant large land mammal. Our affect-laden intuitions and our use of language help us to align effectively not only with people we know, but also with all other members of our species, in playing social games.

The approach I advance here is nested within evolutionary psychology and moral psychology; the second part of the chapter relates the Harmony perspective to recent work in those fields by E.O. Wilson, Christopher Boehm, Jonathan Haidt, and others.[1] The perspective here offers an account of human beings as hypersocial, intuitive, basically ethical animals that can be boiled down to the idea that we as human beings are "addicted to the drama," as the Black Eyed Peas sang and rapped some years ago.[2] Our fascination with our deviations from virtue helps us be virtuous, but it can also blind us—and "us" here includes social scientists, not only the populace. Among other things, it blinds us from taking as seriously as we should a view of ethics as solving social games—something we are very good at as a species—as opposed to viewing ethics in terms of extremely hard-to-realize principles that we constantly fail at, such as Immanuel Kant's principle of following the moral law for its own sake or Jeremy Bentham's principle of maximizing the greatest good for the greatest number.[3]

In a story about a family on a Western car trip, John Updike wrote that America is a vast conspiracy to make you happy.[4] If we switch the object from happiness to social alignment, and broaden the lens from America to our species from its earliest days to the present, we have our core story: Human moral psychology is a vast conspiracy to make us Harmonize.

Game theorists and students of game theory have devoted much thought—decades of it, in my case—to the difficult social games, Disharmony Games, Partial Disharmony Games, and Imperfect Harmony Games, discussed in the introduction and addressed more fully in Chapter Three in this book. In these tricky games, egoistic players are tempted to bully, cheat, shirk, mistrust, and engage in other social crimes and misdemeanors to get ahead and to avoid losing out, and altruistic players face their own difficult, parallel dilemmas. Such attention to difficult, disturbing games is warranted. But the core message of this chapter and the next, and a core message of the book, is that social scientists, and the rest of us, should also devote attention to less dramatic, happy social games. These don't have a standard name, probably because they are not usually conflictual, or tricky to analyze. These are games in which what the players feel and want—which may stem from altruistic or sympathetic motivations just as much as egoistic ones—are aligned.

We will call them Harmony Games. In this chapter, I discuss the basic concept of Harmony. In the next chapter, I refine the concept by relating Harmony Games to the Sanguine, cheerful temperament with which they have a close affinity, and the other three temperaments.

The payoffs in the social games we play in everyday life are usually ambiguous. This means that people can choose to treat their interactions as a happy Harmony Game, rather than as a Disharmony Game, a Partial Disharmony Game, or a Partial Harmony Game. Most of the time, I will maintain, that is exactly what we do. Our emotion-laden moral intuitions, our speech, and our writing all work to align us successfully.

To expand on the point about ambiguity and to put it in modestly more technical terms: Standard game theory usually (though by no means always) assumes that players know their own and other players' payoffs. The Four Temperaments, moral emotions version of game theory advanced in this book, start with the opposite assumption that players do not know their own payoffs or the payoffs of other players. To clarify the assumption: The players may have a great deal of information, and many intuitions to boot, about their own and others' payoffs. But, we assume here, they do not know what their bottom line is, partly because they are subject to multiple moral emotions, or humors, that point in different directions.

The key claim here is that payoff ambiguity is a very good thing for human beings. Human game players who do not know their own, or the other(s)', payoffs are socially expected to treat their interactions in Harmony terms. That social expectation does not apply equally when some, or all, payoffs are known. Because Harmony is a more pleasant game than the other games, and because playing Harmony is facilitated by ignorance of one's own and others' payoffs, we can say that ignorance, in this form, is indeed bliss.

Harmonizing in Practice

As an example of how people seek to Harmonize in practice, consider a subject in a standard lab Prisoner's Dilemma, or Disharmony, game who is told by the experimenter that she and the other player will both get $1 if they both choose Option 1 ("Defect") and $2 if they both choose Option 2 ("Cooperate"). She will get nothing, and the other player will get $3, if she Cooperates and the other player Defects. She will get the $3 maximum, and the other player will get nothing, if she Defects, and the other player Cooperates. In practice, established over decades, through many studies and many thousands of trials, players in the standard low-stakes Prisoner Dilemma lab game play Cooperate around half the time, even though Defect gets them more money no matter what the other player does.[5]

In the Prisoner's Dilemma lab game, both players' dollar payoffs are perfectly known. But, per the Four Temperaments approach to game theory, their payoffs in real life are not known, either to themselves, to their partners playing the game, or to observers. If, as they play the game, they are governed by the self-oriented side of their reactive, calculating Phlegmatic spirit the players will coordinate on treating the dollars as corresponding to their true payoffs. But real as the calm, calculating Phlegmatic part of us is, and real as the self-oriented side is as part of the Phlegmatic and of the three other temperaments, self-oriented egoistic calculation is only a part of one of the four main spirits the players possess. Given the variety of humors in the players, the split within each humor between a more social and a more self-oriented side and the wide variety of social settings, some of which discourage egoistic calculation while others encourage it, the players may well not treat the payoffs as represented by the dollars. In another situation, they may do exactly that because the social frame is more suited to egoistic pecuniary maximization or because of other factors, such as, possibly, momentary biochemical fluctuations between one humor and another in the players.

The Idea of Ethical Focal Points

I learned about game theory, and started worrying about the tension between logic and ethics in certain games, in the seminar taught by Thomas Schelling that I described in the Preface. Professor Schelling gave us many in-class exercises. The one I remember best suggested that people have an intriguing ability to coordinate intuitively on focal points. In that exercise, he asked us the time and place you would go to in New York City to meet up with a person with whom you had made an appointment, but had no way to reach. Nearly all of us coordinated on noon as the time, thus corroborating his idea that human beings can be excellent at intuitive coordination. As I recall it, the winning choice for the location—in a widely split field, with only a minority of us picking it—was Grand Central Station, the answer Schelling mentions in his book, *The Strategy of Conflict*, as the one his Yale students in the 1950s most often gave, logically enough, given that trains from Connecticut to New York City then and now arrive there.[6]

I recently did a modified, multiple-choice version of Professor Schelling's exercise with a business ethics class of Rutgers undergraduates in New Brunswick. As in the survey Schelling gave us, there was no money at stake. Instead, I tried to motivate agreement on a focal point by asking my students to imagine that the people in a very poor village would receive more life-saving public health, such as malaria nets and clean water, the more all of us in New Jersey agreed on one answer to each question. On my version of the New York City question, the fairly close winner in a widely split vote was the waiting room at Penn Station, where the trains from New Brunswick, Newark, and the rest of New Jersey arrive in Manhattan.

Of the eight questions on the survey, I got the best convergence by far on one. In one question, I asked my students to pick one person out of the following five: (1) Abraham Lincoln, (2) Josef Stalin, (3) Kim Il Sung, (4) Adolf Hitler, or (5) Pol Pot. As readers may very well intuit, nearly all of my students picked Lincoln. In other words, they converged on the good guy. In the next question, I asked them to pick one person out of the following, a very different array: (1) Mahatma Gandhi, (2) Martin Luther King, (3) Mother Teresa, (4) Pope John Paul II, or (5) Charles Manson. There was a virtual tie between Gandhi and King (for what it's worth, Rutgers Business School has many Indian-American students), with the two Catholic figures (my school also has many students with Catholic backgrounds) getting a smattering of votes between them. Only one student picked the villain on the list, the California cult leader and mass-murderer Manson. In not converging on the bad guy, my students ignored the "pick the choice most different from other choices" logic that could have theoretically led them to Manson, but did not in practice.

I imagine, and hope, that critically inclined readers will have a number of questions and concerns about the survey results I have just described. I have addressed these in Chapter Five. Here, my aim is to introduce the concept of ethical focal points at a theoretical level. In a nutshell, the idea is that ethical concepts, notably the distinction between good and bad, or good and evil, can be powerful and extremely useful devices—ethical versions of Professor Schelling's focal points—for people to coordinate their expectations and actions in the direction of Harmony in a profoundly ambiguous, uncertain, complex world.

In the moral emotions version of game theory that is central to this book, the basic logic that underlies the claim about people construing reality in Harmony terms relies first on the already stated assumption that human social reality is fundamentally ambiguous: We do not know which game "in reality" we are playing. It also relies on a second key claim, which is that the players do better if they are able to coordinate intuitively on a game. So, for example, if one player acts as though the game they are playing is Harmony, while the other acts as though it is Disharmony, they are in trouble. If they both play the same game, they both do better.

Now, we can return to the core claim: Compared to alternative games, a Harmony game is a compelling ethical focal point for players who do not know the actual game they are playing. Other potential focal points, especially Disharmony, are all attended by serious, unpleasant problems. By contrast, a Harmony game is pleasant. In playing the game of life, humans intuitively coordinate with others in playing Harmony, just as the students in my class intuitively coordinated on Abe Lincoln.

* * *

At the Café

In what follows, I expand on my core contention that humans are very skilled in intuitively aligning with other humans to construe ambiguous realities in terms of Harmony games rather than in terms of other, more unpleasant games. Let's begin by considering a simple, everyday situation: your interaction as a customer with a worker who is preparing your coffee at a café.

If you both reflected on the situation, you and the worker could view your interaction as a Disharmony game in which both of you have an incentive to rip off the other; after all, the worker and the café could likely get an edge over you by stinting on the ingredients, and you could likely get an edge over the store by claiming with calm confidence that you ordered a bigger drink than the worker has prepared for you. Or you could both analyze the game as an Imperfect Harmony game in which mistrust is an issue: the worker might focus on the thought that you won't leave a tip, and you might focus on watching the worker like a hawk. Or you could decide that the game is a Partial Disharmony game in which bullying is an issue: the worker and the store could see the transaction as a chance to push you into ordering more expensive options for your drink, and you could see it as a chance to get the worker and the store to bend over backwards to win your favor and continued patronage.

The claim here is that instead of viewing your interaction in terms of cheating, mistrust, or bullying, you and the worker are likely to coordinate spontaneously and intuitively in treating the social game you are playing as a Harmony game, in which you both have a shared interest in being mutually accommodating in performing your social roles as customer and worker. Whether the game you are playing is really Harmony for both of you or not—perhaps you in fact are in a mistrustful mood this morning, and you have an urge to watch the worker like a hawk—both of you are subject to an expectation that Harmony is the game that you should be playing together, and most likely it is the game the two of you will in fact both play.

If you deviate from Harmony—perhaps you in your mistrustful mood do in fact watch the worker like a hawk—you then feel a further pressure to turn the game into Harmony retrospectively, perhaps by leaving a tip when you usually don't, or by leaving a bigger tip than you usually do. Some of the time, Harmony breaks down, of course, and your interaction with the worker is marred by cheating, mistrust, or bullying, or by a suspicion of them. But the vast majority of the time, the vast human conspiracy works: Harmony prevails.

The Orlando Airport

Human skill in intuitively treating ambiguous situations as Harmony games applies to multiperson interactions as well as to one-on-one interactions. I'll use as an example my fellow passengers and myself in the Orlando airport last August when I was flying back from the Society for Business Ethics conference. My fellow passengers demonstrated the merits of intuitive Harmonizing, while I—classifying the social games we were playing and taking notes—exhibited the limitations of a reflective, calculating approach to understanding social reality.

To get into security, we all had to go through a poorly designed no-rope merge with many people converging in a large human funnel higgledy-piggledy to go into the area with the scanning machines. "Aha!—it's a Disharmony/cheating Game in which we all have an incentive to push ahead while the others cooperate," I thought, feeling a twinge of anger at the inefficient design that led us into the problem, and jotting down a note. I then modified my thought about the game we were playing: "It's also a leadership game—you do want some people to lead by going ahead, rather than having everyone hang back." Meanwhile, though, I and my fellow passengers were all moving ahead more or less effectively. In a matter of minutes there were tens and perhaps hundreds of Harmony Games—or, you could say, one big Harmony Game—with people remaining in place and moving forward in a huge, slow, step-by-step dance.

After some minutes, we got sorted by ropes into lines and approached the security desks. The person in front of me broke a rule by walking in front of a large red stop sign while the security person was checking the ID and boarding pass of the person to my right I did not notice the rule violation at the time, but the security person, who presumably cherished the small island of space afforded by the stop sign, did. He leaned into the microphone in front of him and made an announcement in a low voice that could be heard by the nearest ten of us or so about not stepping in front of the sign until the person before you is done. The person in front of me stepped back. The security person then said in a friendly, shoulder-shrugging tone to him, "You were just a bit in front—gotta follow the rules," and the person nodded.

In the notes I was taking, I classified the interaction between the security person and the rule violator as a leadership, or Imperfect Harmony, game in which the security person prevailed over the violator. From the perspective of detached calculation, it strikes me as a reasonable way to classify their interaction, and from the perspective of normative ethics, I believe that the security person's leading and the violator's following was the overall best, or "Highest Joint Value" outcome for the game. But from the immediate, intuitive perspective of the players, we have something different. We have a success, if a somewhat tension-filled one, in collaborating to treat a situation that could be understood as conflictual as a Harmony Game. Through a well-done performance that combined rule-enforcing firmness with an apologetic tone that made it clear he was acting in role and that minimized the chance of the violator feeling singled out, the security person allowed the violator to join with him in one version—a situation-specific and culturally specific one, to be sure—of the ethical project of Harmonizing.

On the Streets in Lagos

The vast conspiracy of human moral psychology to make us effective in solving social games crosses national, cultural, and historical lines. In a recent novel,[7] Nigerian-American author Teju Cole describes how the body language of people walking on the streets in Lagos needs to be boldly self-confident. In meeting a stranger's eye, Cole's narrator writes, one must convey a "you don't want to mess with me" swagger. On the buses of Lagos, the conductors, or touts, walk with their chests thrust forward, signaling their self-assurance to the passengers.

Human beings being what we are, we will have opinions, perhaps righteous or self-righteous ones, about different styles of human Harmonizing. The point I want to suggest in what follows is that Harmonizing is taking place in Lagos on the bus and the streets described by Cole's narrator, just as it is in the American café and airport I described.

Harmonizing styles are different, to be sure. In Cole's novel, the narrator's well-off family tells him, a visitor from America, that he should beware of black magic on the bus, in the course of trying to deter him from taking a means of transportation used by the poor of Lagos, rather than by affluent people such as themselves. I imagine that people in a reserved Scandinavian-American family in the affluent New Jersey suburb where I live, with a worry about a visiting relative from Oslo who wants to ride the bus to the fairly poor city of Newark a few miles away, would express their concerns differently from the way the people in the narrator's well-off Nigerian family expressed theirs. The reserved Norwegians in my hypothetical, or the slump-shouldered, mild-mannered hipsters in a Brooklyn coffee shop, present themselves differently from the thrust-out, ever-alert Lagosian touts and pedestrians.

Are the narrator's "you don't want to mess with me" Lagosians in fact Harmonizing? In one view, they are not; mistrust, rather than Harmony, rules Lagosian street life. In the view adopted here, they are; the Lagosians are aligning their self-presentation with others' self-presentations in a way that solves games of social interaction on the Lagos streets, just as the alignment of more subdued styles of self-presentation does so in the Norwegian family and the Brooklyn café. Some styles of alignment are more deferential—"together we believe in peace," others more aggressive, or Choleric—"together we believe in peace, and in not letting ourselves get ripped off!" some more reserved, others more emotional, some more oriented toward aligning at the level of individual and social productivity, others at the level of intimate social relations. But all of them allow us to join with others in a social life in which we constantly and routinely accommodate to others and they to us. All of us—touts and hipsters, families in Lagos and in New Jersey and in Norway, moderns and our prehistoric ancestors fifty thousand years ago—are, by nature of our humanity, engaged in a constant, intuitive process of aligning, of creating Harmony Games of different kinds.

Harmonizing, especially in its more intense forms, often involves the Harmonizers creating an us–them community from which an outsider is excluded. In his ride through Lagos, Cole's narrator spots a woman on the bus reading a Michael Ondaatje novel,[8] and feels an immediate, intense bond with her, wondering how she concentrates on the novel's intricate prose with the aggressive, out-thrust tout striding back and forth a few inches away from her in the aisle of the bus. Though one imagines that Cole's intellectual, literary narrator plays Harmony perfectly proficiently with the tout, the intensity of the one-way Harmony Game he plays with the woman is connected, one intuits as a reader, to his placing the nonintellectual tout clearly outside of the Harmony he imagines with his novel-reading peer.

* * *

The Evolution of Moral Psychology: The Move from Pessimistic Rationalism to Optimistic Intuitionism

In the 1970s, the optimistic perspective on human nature advanced here would have been heterodox in the field of moral psychology. At the time, a pessimistic perspective depicting ordinary human nature as ethically deficient was ascendant in the field. Stanley Milgram's famous authority experiments, in which most of his New Haven subjects obeyed a lab-coated experimenter's instructions to shock subjects who were screaming in (simulated) pain in another room, and Philip Zimbardo's prison guard experiment, in which Stanford students fell all too quickly and well into roles as victimizers and victims, provided dramatic accounts of how ordinary people could readily become complicit in evildoing. These outcomes supported a pessimistic sense of how the then-recent horrors of concentration camps, the gulag, and the Holocaust could be rooted in a deeply flawed human nature.[9] Lawrence Kohlberg's theory of stages of moral development provided a conceptual underpinning for moral pessimism. In the scale advanced in the theory, most people were lodged, along with Milgram's obedient subjects, in the less elevated Stages 3 and 4 of getting along with people and obeying social norms, with a relatively small number of people occupying the higher Stage 5 tier of being guided by moral reasoning based on principles, or the elusive Stage 6 moral pinnacle of moral universalism in the spirit of Gandhi or Martin Luther King.

Today, Milgram's and Zimbardo's experiments and Kohlberg's scale[10] remain mainstays in many business ethics classes, including my own. But since the last years of the last millennium, Kohlberg's approach has been on the decline, and a new perspective in moral psychology has been on the rise. This new perspective emphasizes the intuitive, emotional nature of everyday human morality, and jettisons Kohlberg's moral development scale. This perspective in moral psychology, I believe, should also be a mainstay of business ethics classes, and of our general cultural discourse; the Harmony perspective offered here is, in part, a contribution to this new approach.

As with any change in intellectual climate, many factors and many people have contributed to the decline of the old perspective and the rise of the new. Carol Gilligan's feminist criticism of Kohlberg's scale for elevating principled reasoning over sympathetic care played a significant role in weakening

Kohlberg's particular version of rationalism.[11] Alan Fiske, Richard Shweder, and other social scientists played important early roles in laying the groundwork for a new perspective.[12] More recently, Jonathan Haidt and Joseph Henrich have been leading figures in developing the new, relatively optimistic contemporary perspective on the ordinary human morality of intuitive social alignment. Under Haidt's Moral Foundations Theory, values of authority, sacredness, and group loyalty that occupy lower levels in Kohlberg's moral hierarchy are placed alongside, instead of below, values of care, fairness, and freedom.[13]

A striking, large-scale project that helped establish the new perspective perhaps more than any other consisted of Henrich and his coauthors studying fifteen hunter-gatherer societies in the 1990s.[14] They examined how respondents in the Amazon rain forest, the East African savanna, the Australasian archipelago, and elsewhere around the world played "dictator games" in which one player could keep all the money or share with another player and "ultimatum games" in which one player made a take it or leave it offer that the other player could reject. Many previous studies involving Western respondents, mostly college students, had established that players tended to share a considerable proportion of the pot of money, often around 20 to 30 percent, in dictator games, and to offer a larger proportion of the pot, typically around 40 to 50 percent, in ultimatum games. Such studies suggested the importance of altruism in both games, and also indicated the importance of concern about being punished, or about looking bad to others or oneself, if one made a low offer in the ultimatum game. But these interpretations of the studies had been met with skepticism by critics, who wondered whether the results were dependent on the amounts of money at stake being very small for well-off, highly educated Western respondents. By showing that hunter-gatherers for whom the stakes were very large responded in a similar way overall around the globe, giving a considerable share of the pot in dictator games, offering yet more in ultimatum games, and often refusing low offers, Henrich and company provided a nice four-sided portrait of human beings as simultaneously motivated by concerns about their own payoffs, fairness to others, punishing the unfair, and not looking bad.

Introducing the Other Side: "By Nature, We're Not So Good at Solving Social Games"

In such condition there is no place for industry, because the fruit thereof is uncertain, and consequently, not culture of the earth, no navigation, nor the use of commodities that may be imported by sea, no commodious building, no instruments of moving and removing such things as require much force, no knowledge of the face of the earth, no account of time, no arts, no letters, no society, and which is worst of all, continual fear and danger of violent death, and the life of man, solitary, poor, nasty, brutish, and short.

—Thomas Hobbes, *Leviathan*

Against this chapter's optimistic claim about human nature stands a venerable line of religious and secular thought that views us as highly flawed creatures. Many of the many critics of human nature take a rigorous, idealistic view of ethics.[15] These critics of our nature may be perfectly ready to agree with the contention here that people are very good at solving everyday social games, but see that skill as only marginally, or not at all, related to whether we are ethical.

The long-standing debate over the goodness or badness of human nature has less to do, I would suggest, with disagreement on facts—much as the facts on some unclear matters, such as how high the rates of human-on-human violence were in the early days of our species, are relevant—than with different ways in which we understand what ethics is. Under a variety of idealistic, rigorous definitions of ethics, such as Kant's idea doing our moral duty for its own sake rather than because of social pressure, or for that matter because of our own amiable, well-socialized dispositions, we do not fare well.

In riding the train to work recently, I got a small lesson in how idealist rigor about ethics is part of the everyday toolkit of assumptions we use to navigate the world in modern American culture, which one might reasonably think is comparatively practical and relaxed in its notion of ethics, compared to most cultures in human history. A literary neighbor who has written many nonfiction books and who has a nicely inquisitive, Socratic style noticed me walking down the aisle, and motioned to me to come on over. I sat down, he asked me what was going on, and I told him about this book. He asked for specifics, and I told him about the "we're better than we think" thesis of this chapter. He asked me for an example, and I trotted out the focal point game. He looked quizzical: "Yes, but why should that be surprising? People get along, sure . . ." Then he shrugged, and semi-questioned, semi-asserted in a quieter voice, "But their ethics(?!)" We then moved on to talking about the process of writing and publishing a book, with me listening to him as someone who knew far more than I about that subject. The thought for present purposes is simply that my neighbor, unrepresentatively articulate and well published as he is,[16] illustrates, I believe, how rigorous ethical perfectionism—"ethics is the hard stuff, not what most of us do well!"—is a starting point for many of us.

In this section, though, the critics of human nature I want to discuss are philosophers and scientists who operate from a practical, nonrigorous definition of ethics. Thomas Hobbes and Steven Pinker are two good examples of thinkers and writers who, I believe, work from an understanding of ethics that is similar to the solving social games approach proposed in this book. Yet they arrive at more pessimistic conclusions than mine about our original ability as a species to solve social games. We have gotten better, in their account, because of the rise of central government—Hobbes' Leviathan—and of enlightenment values, which in Pinker's simultaneously critical and optimistic account of human nature over time, have led us to become much gentler than we were in what he describes as our shockingly violent past.

Toward the end of his powerful narrative of human history, Pinker suggests why the grim logic of the Prisoner's Dilemma—which he terms the Pacifist's Dilemma—has become less overwhelming over time. Instead of being faced with the full, unmitigated force of the case for individually rational but socially destructive violence that our prehistoric ancestors faced, we now have good alternatives to the Dilemma: "Leviathan," in which aggression is punished by the sovereign; "Gentle Commerce," in which peace in the form of trade and growth becomes much more valuable than destroying the other; "Feminization," in which people do better to act peacefully no matter what the other does; and "The Expanding Circle," in which we come to view the other person's or side's payoffs equally with our own.

Pinker suggests that we can now view reality in terms of games other than the Dilemma, or Disharmony, to employ the term I use. We can see it in terms of what I call Harmony Games—Pinker's Leviathan, Feminization, and Expanding Circle games all have Harmony payoffs, with both players having shared interests. One does not need to see the world in terms of unpleasant Disharmony and Partial Disharmony Games, in which violence and other antisocial conduct pay off for the antisocial.

I strongly agree with Pinker's point, which is also a core idea of this book, that the game one is playing is often unclear, and I am also in agreement with his empirical claim that interpersonal violence was much higher in prehistoric times than it is now, much as the magnitude of the difference is very hard to know. My basic point on behalf of our prehistoric ancestors is narrow but, I believe, important: They faced ambiguous payoffs, just as we do now, and did not truly know their own payoffs, any more than we do now. I believe that our prehistoric ancestors, like us, lived in ethical doubt as to what their reality in a particular situation was.

Did our male ancestors in their heart of hearts genuinely prefer killing another family to gain extra land over peace, as they would if they were truly in Disharmony? Or did they genuinely prefer peace, as they would if they were actually in Harmony, either because killing felt repulsive, or because the benefits from trading and voluntary intermarriage—or simply moving on to new land—felt greater, or for both reasons? Faced with aggression from a neighbor, was our ancestors' true desire to fight back, or to yield (or leave)? I don't think there is a single, determinate answer to these questions that I, Pinker, or for that matter our ancestors themselves could ever arrive at as to their true desires. In particular, it is a mistake, I would suggest, to believe that Disharmony—the Prisoner's Dilemma—is the "real" or "true" description of the payoffs facing our ancestors, as opposed to another game that is less nasty and brutish.

What our ancestors in the earliest days of verbalizing, hypersocial *homo sapiens sapiens* faced, I would suggest, is what we still face: The truth of a species whose members want different things at different times and with different parts of themselves, and who in realizing that truth know that they do not know themselves. Given that fundamental uncertainty, it made sense for our early ancestors, and it makes sense for us, to coordinate intuitively on treating interactions with other members of our species as Harmony Games.

Impelled by the logic of coordinating on Harmony, early verbalizing human beings quite likely adhered then as we do now to a very high degree on the basic idea of virtue—on not cheating, robbing, or killing; on cooperating in following or leading; and on trusting others to trust you.[17] Allowing that their rates of interpersonal violence were much higher than ours, as Pinker persuasively argues, they likely adhered, as we adhere now, to a quite high degree, though certainly an imperfect one, in practicing the basics of virtue. That is the broad story of our past that I would suggest, understanding the deep uncertainties involved, to which I will return shortly, in telling stories about what we were like in our original environment of evolutionary adaptation.

The Harmony perspective advanced here is reconcilable with the hardheaded perspective advanced by Hobbes and Pinker. For intuitive social coordination on a Harmony game to succeed, the Harmony game has to be one that allows people playing in accord with virtue not to be disadvantaged over time in comparison with other people. Human nature needs an array of moral emotions that allow us to solve tricky games, especially Disharmony Games—the nastiest and most difficult—most though not all of the time. Coordinating on Harmony works if and only if Harmonizers can deal effectively with those who misbehave. Wishful thinking—"Let us imagine a blank slate of human nature and act as though it is our reality!"—is not a way to achieve stable Harmony.

To give an example of a key difference between humans and a closely related species, and how that difference enables humans to Harmonize: Jane Goodall tells a harrowing story of a female chimpanzee and her daughter collaborating in attacking another mother chimpanzee and then killing and eating her infant child. Only fifteen minutes or so later, while the killer mother was still feeding on the infant, the mother whose child was killed went over to the killer and reached out her bleeding hand to the killer; the two briefly held hands.[18] A human mother would not have done the same fifteen minutes after her baby was murdered, we may safely assume. In us, there is a powerful punishing anger at wrongdoing that drives us to retaliate against the human equivalents of the killer mother. That deep Choleric desire for righteous vengeance against those who treat social interaction as Disharmony helps us as a species to play Harmony.

The relative optimism here about how well we have been evolved to solve social games from our beginnings as a species is rooted in an approach to human evolution that emphasizes selection for pro-social traits, and that has been gaining strength over the last few decades among scientists who study evolution. To that approach we now turn.

* * *

Social Memes on the March, Selfish Genes in Retreat: The Evolution of Evolutionary Psychology

> There can be no doubt that a tribe including many members who, from possessing in a high degree the spirit of patriotism, fidelity, obedience, courage, and sympathy, were always ready to give aid to each other and to sacrifice themselves for the common good, would be victorious over most other tribes; and this would be natural selection.
>
> —Charles Darwin, *The Descent of Man* (1871)

At its core, the logic of Harmonization is the logic of biological and cultural evolution. In the hardheaded version of Darwinism that is implied in the old phrase "nature red in tooth and claw" and in the phrase "the selfish gene," popularized by Richard Dawkins in the 1970s, the logic of evolution is fundamentally narrow and self-interested. Over the last several decades, this view has been losing ground to socially oriented interpretations of evolution that harken back to Darwin's own belief in evolution as a force fostering cooperation.

One socially oriented perspective views human evolution in terms of socially transmitted "memes" (another Dawkins coinage)—ideas and behaviors inscribed into our genes, our cultures, and their interactions. In another interpretation—the one that will be emphasized here and that has been pioneered by entomologist and evolutionary theorist E. O. Wilson[19]—humans and the social insects have become the planet's most successful species because in us evolution has fostered intensely social, or hypersocial, to use Wilson's term, behavior.

In addition to E.O. Wilson, many others in a variety of fields, including anthropologist Christopher Boehm, biologists Elliott Sober and David Sloan Wilson, and economists Samuel Bowles and Herbert Gintis, have played important supporting roles in the now decades-long turn toward viewing human evolution in terms of social behavior.[20] Boehm's work, with its emphasis on selection for pro-social traits within groups—hunter-gatherers punish their antisocial peers, and resist bullying leaders—is particularly relevant for our purposes.

For many years, evolutionary theorists who were interested in pro-social behavior focused on group selection, as Darwin himself did. That focus on group selection, which continues to the present in the socially oriented Darwinism of David Sloan Wilson, relies on Dilemma models that assume ethical behavior is disadvantageous to individuals who practice it, while at the same time being beneficial for their groups.[21] Under the definition of ethics here, which views it in terms of the stable, sustainable solving of social games, the view of ethics as individually disadvantageous misses the essence of human nature, much as it resonates with certain aspects of our lives.

Ethics in the understanding advanced here is generally, though not always, advantageous to individuals in their relations within groups, not simply a beneficial quality for their groups in relation to other groups. The approach here thus aligns with Boehm's view that conscience evolved as part of a process of social selection within groups, and with the broad definition of pro-sociality that he, Bowles, and Gintis, adopt, as opposed to a narrower definition of pro-sociality as altruism that is disadvantageous to the altruist.

The E. O. Wilson–Boehm–Bowles–Gintis version of modern Darwinism that is adopted here by no means proclaims the reign of sweetness and light. Always, as Wilson has noted, there is ambiguity and ambivalence in human ethics. One's own enjoyment is certainly one valid moral *desideratum*—but what about the moral claims of the worker in the coffee shop, of one's family, of one's firm, of one's town, of one's nation, and of one's species, or for that matter of all sentient life, of all life, of all being, or of all nonbeing? The position taken here that human beings have evolved biologically and culturally to be ethical in the sense of being very good at solving social games with other human beings should not be confused with the position that we have evolved to be happy, serene, or saintly.

Hypersociality in ants, bees, wasps, and termites has everything to do with tightly programmed social routines, and very little if anything to do with social emotions, benevolent or otherwise. The much less tightly programmed hypersociality of human beings is deeply linked to social emotions, but not necessarily positive ones. In particular, human hypersociality as theorized by Boehm, Bowles, and Gintis requires that anger at others' deviations from norms, vengefulness, and punishment be central parts of our makeup. Bullies and would-be tyrants are kept in line, in Boehm's account, by a reversed hierarchy of egalitarianism, in which resentful anger is a key tool in resisting antisocial predation.[22] Our troubling capability for rage and vengeance makes our often self-interested "false positive" errors in identifying other individuals and groups as amoral or immoral much more dangerous than otherwise—but at the same time, it prevents our hypersociality from being destroyed by asociality, or by antisociality.

The perspective adopted here on the nature of our earliest ancestors—pleasure-seeking and pain-avoiding, angry, ashamed, sympathetic, and all in all much like us—is mostly derivative of the accounts of Boehm, Bowles, and Gintis. To the extent the account here is original or different in asserting that humans are better from our origins to the present than we often think we are, it draws on two main facts about our history as a species as a basis for a conjecture about our past and present human nature.

The first key fact about us is that we have been extremely successful in multiplying our numbers from a very small base in Africa somewhat over 100,000 years ago, and from another base that was likely very small when some of our ancestors left Africa to populate the rest of the world around 60,000 years ago.[23] The second key fact is that compared to the recent period when humans have become by a very large margin the world's dominant large land mammal, and have occupied virtually all of the world's habitable land, our environment of evolutionary adaptation featured a paucity of humans, and an abundance of dangers and opportunities associated with other animals, including other hominids, and with nature.

My conjecture about human nature is that from very early on, *homo sapiens sapiens* has had an intuitive, gut-level "us" feeling about other humans relative to the rest of the world that has contributed significantly to our skills in playing Harmony Games with other humans, to differentiating ourselves from other primates such as chimpanzees and bonobos, and to our rapid rise from insignificance to a substantial role in the affairs of the planet. Our in-group feeling about our own species was not an unmitigated good in our environment of evolutionary adaptation, and it is certainly not an unmitigated good now. It was not a good thing for woolly mammoths, saber tooth tigers, chimpanzees, and Neanderthals to be out-groups for our ancestors, and it is not a good thing now for the species already gone or threatened by a new wave of anthropic extinctions. But human feeling for other human beings was, and is, a highly valuable, deeply worthwhile part of our nature.

It is good that our modern moral discourse includes the utilitarian philosopher Peter Singer denouncing our preferential sympathy for other human beings as "species-ism"—self-critical energy on behalf of animals and nature is worthy. But, I believe, it is a far, far better thing, in fact an invaluable one, that we have a "we" feeling about our species as one key component of our nature.[24] Similarly, it is good that we have the ever quotable, ever provocative Dawkins questioning our self-love by noting that human fetuses with minimal cognitive and emotional capacities receive more protection than chimpanzees with considerable reasoning skills and highly developed emotions.[25] But it is wonderful, I would suggest, that our complex, ambivalent human nature includes an element of home team feeling about all other human beings. That spark of connection between all of us as human beings has enabled our success as a species and helps make our lives worth living (Figure 1.1).

	Player 2	
Player 1	**Unknown payoffs**	Known payoffs
Unknown payoffs	Harmony!	Not so good . . .
Known payoffs	Not so good . . .	Not so good . . .

Game players who do not know their own payoffs, or other people's payoffs, are socially expected to create Harmony Games. That expectation does not apply as strongly when payoffs are known. Because Harmony is a more felicitous game than Disharmony and other games, and because creating Harmony is facilitated by ignorance of one's own and other people's "true" payoffs, we can say that ignorance is bliss.

Figure 1.1 Ignorance Is Bliss

Summary

In our interactions with other people, with rare exceptions, we do not know the payoffs that we, or others, will receive for given combinations of actions. This fundamental ambiguity allows us to coordinate on creating and playing games in which we have shared feelings and interests, which may be altruistic, or social, as well as egoistic, or individual. These games, which have been little studied in classical game theory, may be called Harmony Games. Human nature may be usefully understood as a universal conspiracy to Harmonize intuitively with other humans.

The strong Harmonizing quality in human beings is in part species-ist, and it has been much less good for other creatures, such as mammoths, than it has been for us. All in all, though, a Harmony perspective of humans as brilliant intuitive creators and players of social games offers a basically positive perspective on our nature, in accord with recent trends in evolutionary psychology and moral psychology.

Exercises

1. (a) Read aloud and discuss, or reflect on, passages from Updike's story and Cole's novel, with a focus on their temperaments as well as on the societies they portray; (b) Use Tolstoy's "All happy families are alike . . ." quote as a starting point for discussion, or reflection, on the simultaneous human desires for happiness and for drama.
2. (a) Watch and discuss, or reflect on, videos showing CEOs and their deputies with similar styles of dress and/or expression; Apple executives Steve Jobs and Tim Cook are a good example; (b) Read and discuss, or reflect on, a profile of E. O. Wilson that focuses on his having water poured on him by protesters yelling, "Wilson, you're all wet!"
3. (a) Watch and discuss, or reflecting on, all or part of videos by Jonathan Haidt, Joshua Greene, and Joseph Henrich on human moral psychology; (b) Read and discuss, or reflect on, online material on William H. Whyte's Public Spaces Project and how humans enjoy crowds and social diversity within crowds.
4. (a) Do the quiz in the endnotes (and discussed further in Chapter Five) on ethical focal points; (b) Read and discuss, or reflect on, the passage from Jane Goodall on the chimpanzees of Gombe in which she describes the killing and eating of the infant daughter of Melissa by Passion and her daughter Pom.

CHAPTER 2

The Harmony Games

The central aim of this chapter is to relate the four temperaments to the Harmony perspective on human nature that was introduced in the opening chapter. I suggest that human life consists in substantial part of quicksilver transitions from Harmony in one mood to Harmony in another mood. In what follows, I propose an eightfold division of Harmony: four temperaments times two forms of thinking/feeling. Each temperament can Harmonize in an intuitive System 1 mode or in a reasoning System 2 mode. In that eightfold array of Harmony, I identify one weak link. Modern humans are strong, I contend, in creating seven kinds of Harmony Games but are much less effective in creating one: Sanguine System 2 Harmony. As much as we believe in the Sanguine as the pinnacle of our aspirations, we do not intellectually respect self-help books, and other guides to happiness, even as we do respect calculating, compliance-oriented, and argumentative modes of reason.

In the second part of the chapter, I turn to normative concerns associated with Harmony. Given the ubiquity of human Harmonizing, Harmony Games are nearly always part of serious moral wrong, and sometimes help such wrong to occur. Notwithstanding that important problem, I argue that Harmony and the drive to create it are good. Further, a view of humans as basically good, Harmonizing creatures is itself preferable, I argue, not only on factual grounds, but also on ethical grounds, to a view of our species as morally defective.

Eightfold Harmony

Harmony is the simplest type of social game, but it has its own considerable complexities. One is that Harmony, like other social games, is indeterminate. Our success as a species lies in our ability to create Harmony, not simply to discover it in a clear-cut, preset form. A related complication is that whether a given game is Harmony or a difficult one is relative to the emotional and ethical makeup of the players of the game. A game that is Harmony for two egoistic players may not be for two altruistic players, and the other way around, as will be discussed in the next chapter.

A basic conjecture in this book is that the classical system of temperaments can be applied to identify four important ways we Harmonize with other people, each of which is divided into an intuitive System 1 mode and a reasoning System 2 mode. In this section, I give examples of each of the eight versions of Harmony games: Sanguine System 1 and 2, Phlegmatic System 1 and 2, Choleric System 1 and 2, and Melancholy System 1 and 2.

First, let's consider Choleric Harmony, and its important competitive subtype. In this version of Harmony, we align with others through competition, through pride, and, sometimes, through anger and indignation. For examples of the intuitive System 1 mode of Choleric/Competitive Harmony, think of children or adults playing games like Rock–Paper–Scissors, four square, checkers, or soccer with one another. They are trying to win—but in these and all other competitive games, the players are also coordinating with one another. Styles of Choleric/Competitive Harmony differ. One group of children will play quietly with never a sharp word or an outburst, another boisterously; in one soccer league, players will be poker-faced, in another operatic.[1] But in all these cases, Choleric/Competitive Harmony is the rule, the normative expectation. A player who feels misaligned with other players—as, say, a soccer player used to emoting, or to being poker-faced, might feel in going to a new league in which the opposite mode is the norm—will recognize a pressure to join in the prevailing mode of Choleric/Competitive Harmony in the social milieu.

For an initial example of the reasoning System 2 mode of Choleric/Competitive Harmony, think of two siblings arguing more, or less, civilly with one another—or appealing to their parents—on whether the sibling who owns a toy should be able to get it back from the one who is playing with it. For an adult equivalent, think of lawyers for the plaintiff and the defendant engaging in oral argument before a panel of judges. For a third, extremely impressive, example of Choleric/Competitive Harmony and its successful alignment of disagreement, think of the simultaneously deeply passionate and deeply civil 2014 referendum on independence for Scotland. For a final, close to home, example, imagine an ethics colloquium in which one speaker energetically advances the position that System 2 moral reasoning rightly applied supports a liberal, universalist worldview, while another speaker passionately advances the position that the advocates of liberal universalism are themselves falling into us–them parochialism.[2]

In System 2 Choleric/Competitive Harmony, the "I'm right, you're wrong!" arguments of the competitors coexist with an expectation that they align with respect to one another in the interest of a social whole of which they are part. This expectation of Harmony may be upheld externally, through means such as the approval or disapproval of parents in the case of the arguing children, or of judges in the case of the arguing lawyers. But it is centrally enforced by the players themselves. In Competitive Harmony, one wants to be simultaneously effective in advocacy and in accommodation. One also has, as in the other types of Harmony, a feeling that it is irrational—weird and strange—as well as wrong not to align with others in playing the game. In Competitive Harmony, one does not feel like Ferdinand, the bull in the children's story who only wanted to smell the flowers and not to fight in the ring. One wants to compete, and one is unhappy if the other player is a misaligned Ferdinand.

With Competitive Harmony, there is often no clear line between the intuitive System 1 and reasoned System 2 versions. If game players are young or immature, their Competitive Harmony may slip back and forth from System 1 to System 2, with disputes over whether one or both sides are gaming or abusing the rules in one way or another. The same is true for players at the top of their games, such as World Cup athletes, coaches, and referees.

The second major type of Harmony proposed here is Melancholy, or Compliant, Harmony. In this mode, we align with others in sadness, and sometimes in shame, guilt, anxiety, boredom, pain, fear, or even horror, in a spirit of shared obedience to social norms. Sometimes these norms are familiar and time-honored; sometimes they are norms we did not realize we shared until a catalyzing moment. Shared yawns, downcast looks at a funeral, or a wince on hearing a friend's bad news are familiar examples of intuitive Melancholy Harmony. Another example from some years back, which like all Harmony games passed in favor of new ones, but which can come back like all the games as Harmony recollected, especially for those of us who were present at the time: People walking in public places in Manhattan in late September 2001, surrounded by hand-made signs with pictures of the missing, affixed to walls and posts.

How about reasoning System 2 Melancholy Harmony? When people are joined by a shared sense of sadness and shame, the argumentative spirit of Choleric Harmony yields to a very different kind of reasoning human spirit that tries to figure out how to obey together. So, for example, in my home state of New Jersey, the Democratic-controlled legislature and the famously pugnacious Republican governor, Chris Christie, moved from their default mode of Choleric Harmony—often difficult to distinguish from Disharmony—into a very different mode after the suicide of gay Rutgers student Tyler Clementi, which led them to join together in supporting and passing complex new law and regulations dealing with harassment, intimidation, and bullying.[3]

In the third major type, Phlegmatic, or Pragmatic, Harmony, we are joined in a calm, practical spirit. As one example of intuitive System 1 Phlegmatic Harmony, think of the overwhelmingly, though not invariably, aligned movements of multiton vehicles and their human operators on roadways, or of pedestrians on a crowded city sidewalk. For another, think of a modern cafe, with muted conversations and most patrons on their laptops or texting. For a third, think of our prehistoric ancestors, working together effectively to bring down prey that was far bigger and stronger than they.

Like Choleric and Melancholy Harmony, Phlegmatic Harmony lends itself to effort-laden reason as well as to automatic intuition. In reasoning System 2 Phlegmatic Harmony, one central role for the Harmonizers is that of calculator, whether as a modern operations researcher or as an ancient accountant. Another central role for Phlegmatic Harmonizers—an especially relevant one given the focus on business ethics in this book—is that of manager. Whether one is a contemporary American executive trying to bring together diverse organizational cultures after a takeover, or an administrator five thousand years ago in one of the world's first cities in Mesopotamia trying to collect and distribute grain, one is engaged with others in Phlegmatic Harmony.

The last proposed major type of Harmony, Sanguine Harmony, is the type that more than any of the others defines the concept. Together, we are happy, flourishing, loving, joyous, laughing, friendly, smiling, cheerful, and/or optimistic. That shared positive experience may be a fleeting one—the Sanguine temperament is proverbially a mutable one—but it epitomizes what most of us mean by Harmony. Intuitive happy Harmony—lovers holding hands, a happy crowd at a parade—is the stock-in-trade of our hopes and dreams. We who aspire to critical consciousness of our selves and our culture may find standard examples of it jejune, if we wish, but we no less than others seek it.

In the pursuit of happy Harmony, all of us, critics or not, have an interesting problem. Competitive Harmony, Compliant Harmony, and Pragmatic Harmony are all readily furthered by effort, by different kinds of reasoning, and by different kinds of social processes and institutions, such as competitive politics, obedience-enforcing law, and efficiency-oriented business. With happy Harmony, the situation is different. No obvious forms of reasoning, and no obvious social processes or institutions parallel in power to politics, law, or business, offer us a reflective, System 2 path to human happiness. In particular, education, psychology, and management do not now, I believe, offer Sanguine logics and ethics that are of comparable power to the valuable but also partial logics and the ethics of Competitive, Compliant, and Pragmatic Harmony. One may conjecture that that lack is a feature rather than a bug in our nature, built into a logic of evolutionary effectiveness that militates against our being satisfied.[4] Still, one may also wish as teachers, or as psychologists, or as managers, for ideas on how we might do better in creating reasoned versions of Sanguine Harmony. Whether or not reasoned paths to happy flourishing exist, and whether or not we can find them, the value of the quest would seem to be considerable.

A final note: We need to bear in mind that Harmony as defined here is an interactive game, in which both or all players have a dominant strategy that accords with their feelings and desires, which may be altruistic, egoistic, competitive, or norm-following. In all of the versions of Harmony just discussed, both players (or all players) would say, if they were to reflect on their alignment: "I am very glad that both of us (all of us) are competing/obeying/working/enjoying in a good way—but regardless of what the other player does (or other players do), that way is what is best for me."

Mutable Harmony: Fictional Portrayals

In this section, I aim to consider Harmony games at a micro, second-by-second, minute-by-minute level. Our lives can be broken down, I believe, into repeated, frequent variations on the four major kinds of Harmony, in their System 1 and System 2 forms. The thought here is that the best source for testing that proposition is fiction, which can be understood as a moment-by-moment ethnography of human life.

Certain writers demonstrate a power to Harmonize with us through their descriptions of fleeting, momentary intuitions, emotions, and thoughts. A compelling story must Harmonize its readers, or hearers, by including Disharmony. If instead of *The Hunger Games*, with its disturbing story of teenagers competing to the death, Suzanne Collins had written stories of satisfied, always happy young people playing badminton without a net, her readership would have been small, we may safely assume. But in addition to Disharmony, I would suggest that an effective story also needs recurrent illustrations of Sanguine Harmony, Phlegmatic Harmony, Melancholy Harmony, and Choleric Harmony. If that is so, a further conclusion is suggested: a rapidly fleeting parade of the temperaments appears in effective fiction because it corresponds to us, and to our moment-by-moment stream of intuitions, feelings, and thoughts.

For an informal test of the idea, let's consider the beginning of *The Hunger Games*. If my conjecture is right, we ought to be able to find instances of the four proposed major types of harmony in fairly short order. In the book's first paragraph, Collins sets the stage on her drama of sixteen-year-old Katniss Everdeen, a fierce girl from the backwoods of Panem. Katniss tells us that her bed is cold and rough and that her sister Prim has had bad dreams and abandoned her to sleep with their mother. We are in the domain of Melancholy, with a shared, aligning Harmony element that is subtler but also central. Katniss explains her sister's bad dreams by saying that this is the day of the reaping. With this odd, unexplained phrase, she creates in her readers a sense of anxiety that draws us into Melancholy alignment with her and her sister.

In the next paragraph, Collins through Katniss moves rapidly from the Melancholy Harmony of the first paragraph to Sanguine, Happy Harmony, and then back again. The paragraph begins by evoking a Happy Harmony of cocooned beauty in Katniss's mother and Prim sleeping peacefully, cheek to cheek, and creating a parallel Harmony in the reader. It then turns quickly back to Melancholy Harmony by asserting the transitoriness of peace and beauty: Soon, the cocooned pair will have to awake; Prim's loveliness will one day fade, even as her once-beautiful mother's already has.

In the next several sentences, the mood shifts to Competitive Harmony, with Katniss describing her relationship to Prim's cat, which she tried to drown years ago when the family was short on food. In Katniss's description of how she now feeds the cat entrails from the kills she make and how the cat no longer hisses at her, the Choleric, competitive element is clear: We are in the domain of humans and animals fighting, killing, and eviscerating. The Harmony element is also present: Katniss and Buttercup are a matched pair of tough customers, Kat and cat, who respect the other and whose shared fierceness benefits the family. Competitive Harmony is also created though the reader's intuitive recognition of a stock, archetypal, drama of sibling rivalry; Katniss's attempted drowning of the cat, we intuit without conscious reflection, represents the deflection of her competitive, hostile feelings toward her beautiful younger sister.

In the next few sentences, which close the first scene of the book, Katniss transitions to Phlegmatic, pragmatic Harmony, with a touch of the Sanguine, as she describes putting on her flexible, comfortable hunting boots, tucking her braid into a cap, and putting in her pocket a goat cheese wrapped in basil leaves that her considerate sister has left for her. As before, Collins via Katniss creates a Harmony game with readers in the course of describing Katniss's Phlegmatic Harmony games; one intuitively feels oneself relating to the material world, and to people, in the practical project of getting dressed and leaving one's house, with—if one is lucky—Phlegmatic Harmony joined in one's own life by a version of the Sanguine Harmony shown in Prim's thoughtfulness and Katniss's appreciation.

So, in a barely more than one-page scene, *The Hunger Games* works efficiently through the four core temperaments and types of Harmony that are treated in this book. In the next paragraph, with Katniss outdoors, the wheel turns from Phlegmatic Harmony back toward Melancholy Harmony, with a description of sunken-faced, broken-nailed miners in District 12 going to work. Quickly and efficiently, Collins through Katniss has taken us through a series of repeated, brief Harmony games, played by Katniss with other people and with inanimate objects, and with her readers, feeling the emotions evoked by her narrative.

How general are the lessons of *The Hunger Games*? Are we like Katniss, shifting rapidly from mood to mood, Harmonizing in different ways with people around us and with the material world? I believe so, much as the usual researcher's line that more study is called for most certainly applies here.

One reaction that I suspect some readers of this book will have is that Katniss, compelling as she is for millions of readers, and much as she may illustrate well the intuitive System 1 types of Harmonization, does not work well in evoking reasoning System 2 Harmony Games. I believe the concern is a fair one. To understand System 2 forms of Harmony, it helps to move from the domain of popular fiction to that of high art. For the greater part of us that is System 1, Katniss works very well indeed. For the lesser, but also important, part of us that is System 2, it will help to move from the muddy fields of Panem on which Katniss fights and triumphs to the empyrean of the world's finest literature.

Do we see the four humors of Harmony represented quickly, or at all, in Henry James, the Master of complex, qualified, self-canceling descriptions of elusive, evanescent events and nonevents? Let's see; the James work I'll examine is *The Tragic Muse*, an 1890 novel in which the central character, Nicholas Dormer, is a married politician who sacrifices his seat in Parliament, his wife, and his family to become a painter. Here is the novel's opening sentence, which evokes us–them Disharmony:

> The people of France have made it no secret that those of England, as a general thing, are, to their perception, an unexpressive and speechless race, perpendicular and unsociable, unaddicted to enriching any bareness of contact with verbal or other embroidery.

A few paragraphs later, James describes a group of four people sitting silently together on a bench at an outdoor art exhibition in Paris:

> They had about them the indefinable professional air of the British traveller abroad; that air of preparation for exposure, material and moral, which is so oddly combined with the serene revelation of security and of persistence, and which excites according to individual susceptibility, the ire or admiration of foreign communities. They were the more unmistakable as they illustrated very favorably the energetic race to which they had the honor to belong.

James in this passage is creating a Phlegmatic, Pragmatic Harmony with his readers. The four travelers are aligned with one another, and with the serene energy of the group of people to which they belong. The us–them tension that dominates the opening sentence has yielded to a sense of a distinctive British style that engenders Harmony, as well as Disharmony, with other groups. In his depiction of Phlegmatic Harmony, James creates it with his readers; together, we experience the calm persistence he describes. In the alignment the passage creates, there is a mix of intuitive System 1 affect and System 2 thought; James invites us to feel both the odd combination of British openness and security, and a reflective sense that the odd combination is not so odd at all.

The focus moves in closer still on the second page, in which James describes the mood that unites the four travelers:

> "En v'la des abrutis!" more than one of their fellow gazers might have been heard to exclaim; and certain it is that there was something depressed and discouraged in this interesting group, who sat looking vaguely before them, not noticing the life of the place, somewhat as if each had a private anxiety.

Moods must change in stories as in life; we have moved now to Melancholy Harmony, to a shared worried, downcast spirit in the four, and in us.

In the same paragraph, James moves his camera in closer, and describes each of the four people, who we learn are a family. The first to sit for her portrait is the mother:

> This competent matron, acquainted evidently with grief but not weakened by it, had . . . a tendency to throw back her head and carry it well above her, as if to disengage it from the possible entanglements of the rest of her person.

The transition from Melancholy to pride and Choler that is suggested in this sentence in carried through in the following one:

> If you had seen her walk you would have perceived that she trod the earth in a manner suggesting that in a world where she had long since discovered that one couldn't have one's own way, one could never tell what annoying aggression might take place, so that it was well, from hour to hour, to save what one could.

We have now entered the domain of Choleric Harmony. This type, like the others, creates its match in the reader, but with a tension that is distinctive to Choler: we feel Lady Agnes's reactive pride, and at the same time react to it and to its manifestation in ourselves. We are invited by James, in a reflective System 2 spirit, to question Choler.

The last of our four proposed major types of Harmony, the Sanguine, appears in the same long paragraph. James portrays the mother's son, Nicholas Dormer:

> I cannot describe him better than by saying that he was the sort of young Englishman who looks particularly well abroad, and whose general aspect—his inches, his limbs, his friendly eyes, the modulation of his voice, the cleanness of his flesh-tints and the fashion of his garments—excites on the part of those who encounter him in far countries on the grounds of common speech a delightful sympathy of race.

We are now in happy, Sanguine Harmony with the pleasant son and his pleased compatriots. At the same time, and more acutely than with the other forms of Harmony we have examined in James, there is System 2 ambivalence here. As modern readers we perhaps feel a twinge of Disharmony at "the delightful sympathy of race," along with a counter-twinge to restore Harmony by reminding ourselves of the way in which James and his contemporaries referred to race where we would eschew the term in favor of nationality, and a possible counter-counter-twinge that the highly reflective James himself likely wanted his readers to feel a twinge of Disharmony at overly facile versions of Sanguine Harmony.

Our conclusion: The succession of humors in Suzanne Collins can also be discerned in Henry James. In only three pages, the prolix Master takes us from his opening statement of Disharmony (which later in the novel he broadens into sketches of Competitive Harmony and other kinds of Harmony in personal Anglo-French relations) through brief depictions of Phlegmatic, Melancholy, Choleric, and Sanguine Harmony. Henry James and Suzanne Collins are very different story-tellers—but both are akin in their mastery of what I am suggesting is the fundamental need of the story-teller to describe and create recurrent, rapidly shifting forms of Harmony games. His work, as well as hers, can be employed to support the proposition that our lives, to the extent they mirror our stories, are made up of mutable Harmonizing.

At the same time, there is a significant System 1–System 2 difference between the two writers. James does not work through the Harmony types as rapidly and fluidly as Collins does, or with her smooth transitions. With him, one feels one is listening to a brilliant, erudite, fussy older relative commenting in staccato fashion on a series of photographs, while with her one feels as though one is watching a movie. His version of Harmonizing engages our reasoning System 2 capabilities far more powerfully than hers does, while hers engages our intuitive System 1 capabilities more effectively.

The difference between Collins and James is particularly marked on the Sanguine dimension. James would not, one feels, ever permit himself a sentence like Collins's "Prim's face is as fresh as a raindrop, as lovely as the primrose for which she was named," at least not without careful System 2 qualifications that would signal his unwillingness to be sentimental. For his intensely, intricately reasoning sensibility, a simple Sanguine spirit will not do, and a complex, reasoned, Sanguine spirit is not readily at hand.

Creating Harmony Games

If the perspective here on eightfold Harmony and moment-by-moment Harmony is right, it has, I believe, a clear normative implication for business ethics, as well as for general ethics. As managers, leaders, followers, friends, citizens, and so on, we are in a position analogous to that occupied by Suzanne Collins and Henry James as creative artists. Whether our gifts are intuitive, reflective, or, as it may be hoped, a blend of both, we have constant opportunities to create Harmony in our daily interactions with other people, and also in our interactions with things, such as animals, machines, and organizations. We may try too hard to do so, like the maladroit Mr. Collins in *Pride and Prejudice*, and in that overexertion, or in other ways, fail. But our effort, I would suggest, is good, as are the varieties of Harmony we attempt to create.

In the next part of this chapter, and again in Chapter Four, I consider morally disturbing concomitants of Harmony. Here, though, I want to stress the basically positive, fortunate quality of the human pursuit of Harmony. As applied to business ethics and to business, the point I am making is akin to, but slightly different from the widely disseminated message of searching for win-win solutions. Win-win, much as it is akin to Harmony as a concept, implies that both sides are negotiating or otherwise acting on behalf of their egoistic interests, and are giving up something of value to get something else of greater value. In that sense, both players in a win-win game are arriving at their second-best, rather than their first-best, outcomes. In Harmony, by contrast, the outcomes may be first-best for both players, and the range of motivations, though it may definitely include egoism, is not limited to it.

We can use the management styles of Herb Kelleher of Southwest Air and Jack Welch of GE to illustrate the difference between Harmony and win-win. First, Kelleher: One could view the "loyalty plus flexibility" approach to employment relations utilized by Kelleher, in which the firm worked hard not to terminate employees, and employees were expected to be ready and willing to perform any job for the company,[5] in terms of a win-win trade-off, in which both Kelleher and the employees settled for less than their optimum outcomes. It is true enough that if he could have gotten complete flexibility in termination, along with total employee commitment and flexibility, Kelleher as CEO would have done better in immediate, egoistic terms, and similarly true that the employees for their part would have done better in immediate, egoistic terms with the reverse arrangement. But in the broader lens of Harmony, one can view the Southwest culture as one in which both Kelleher and the employees achieved what they wanted most.

A similar point applies to the very different "loyalty is over" approach to employment relations associated with Welch. One could view the GE culture as a trade-off, in which Welch and employees both gave up something—employee loyalty for Welch, the company's loyalty for the employees—to get other things they valued more. But one can instead view what Welch achieved culturally at GE in terms of Harmony, a more Choleric, competitive Harmony to be sure[6] than the more Sanguine kind at Southwest, but Harmony nonetheless, with employees and Welch achieving what they truly wanted.

Harmonizing under Disturbing Circumstances

In this section, I consider the challenge, as well as the support, that the ubiquity of Harmonizing presents for the basically optimistic perspective on people advanced in this book. People not only treat ambiguous situations as Harmony in the relatively morally tranquil domains of the coffee shop and the airport. We also create Harmony games with other people in situations that are anything but morally tranquil—the doctor reassures the worried patient with a placebo; Milgram's subjects in New Haven obey the man in the white coat and "shock" the subject; the Ptolemaic queen and her young brother marry; children join together in the ritual eating of the flesh of a deceased parent; and the slaves on the auction block turn around to be inspected by potential buyers.

Harmonizing continues even in the shadow of death. Soldiers follow orders and march ahead resolutely to the front to face the fire of their foes; the condemned prisoner sits quietly while being strapped into the execution chair. A roomful of people who have been transported by train far from their homes to a camp in the Silesian countryside remove their clothes calmly, in preparation for the showers that their guards have told them they will be taking shortly.

These examples of Harmonizing raise questions for the optimistic perspective advanced here on human nature. Can our intuitive excellence in aligning with other human beings socially, impressive as it is, be properly treated as ethical at all, much less as the key piece of evidence supporting the proposition that our nature is basically good? Given that Harmonization can be based on falsity, and that it can serve evil as well as good purposes, is it better understood as ethically neutral, or even ethically negative, rather than as ethically good? Do we not need to defend an intuitive, social account of ethics such as the one given here with moral reasoning?

I would say yes to the last question, which I address in what lies ahead. To the earlier questions, I would respond as follows: First, let us beware of, and not fall prey to, negativity bias. True, that bias has understandable and sensible roots. It is good that we are evolved so as not be blasé about bad things that could harm us—an organism that feels itself to be threatened is an organism more likely to survive. It is also good that we are evolved so that we are highly attuned to the disturbing side of social life—a social organism that has a devotion to drama is an organism more likely to learn from its mistakes and to reproduce itself and its culture. But much as overemphasizing the negative is understandable and in some respects desirable, it should not be yielded to in evaluating our species, our drive to Harmonize, and our modes of Harmonization.

Human beings individually and collectively sometimes do terrible things, and often do bad things. Harmonization can certainly serve malign purposes. But it is a mistake to allow our view of ourselves as a species to be dominated by terrible or bad actions, and it is a particular mistake to view the seeking of Harmony in human interaction as itself malign based on the instances in which Harmony goes awry. Whether it is the good or not, the human pursuit of Harmony is good.

It is part of our basically good nature to be disturbed at how we sometimes do terrible things, and at how our Harmonizing can go awry and become co-opted into wrong. What makes us good is not only that we have intuitive skills in aligning ourselves with other human beings in the constant microinteractions of everyday life. It is also that we are impelled by powerful moral emotions of righteousness, anger, shame, and guilt, associated with the Choleric and Melancholy temperaments. These emotions lead us to be bothered, if sometimes too little and other times too much, by violence, cheating, bullying, defiance, disloyalty, impurity, and the other characteristics of false as opposed to true Harmony games. The combination of our intuitive Harmonizing and our deep emotional unhappiness at Disharmony—the combination of the human bright side of sociability, serenity, and sympathy and the boiling, freezing human dark side of anger, shame, and guilt—makes our ethical natures better than either our bright side or our dark side emotional qualities by themselves would.

Further, I would suggest that it is an important part of our basically good nature to rationalize and refine our moral emotions through reasoning. Reasoning in the view advanced here is a particular type of Harmonization.

In the dialogue of reason, the speaker, writer, listener, and reader, much as we may disagree with one another and indeed are expected to do so, do so in the stylized context of a Harmony game in which we as the players align intuitively in a shared commitment to reason.

With the rise of writing—dramatically illustrated in the contrast between the arguing Socrates and the lecturing Aristotle, with Plato as a bridge between the cultures of speech and writing—there is no direct, emotion-laden, two-way communication between a speaker and a listener, and Harmonization of the traditional kind becomes difficult. Based on our biologically and culturally evolved humanity, nearly all of us can Harmonize decently enough in conversation, but Harmonizing as a writer with an invisible audience of readers is harder—as I can attest to from my own book-writing experience! At the same time, Harmonization remains as central in reasoning through writing as it is in reasoning through speech. Without a real or imaginary other with which to align and be aligned by, reason is silent, still, and dead.

The game of reason is a serious one, and any particular playing of the game may well be anything but gentle and uncompetitive, and may feature a winner and a loser—for example, we on the "we're basically good" side of the human nature debate may marshal feeble arguments, and be routed from the field by the more acute arguments of the "we're not so good" side. But our playing the game of reason together rather than some other, less pleasant game is yet another example of the human genius for construing ambiguous reality in Harmony terms. Reason, rightly understood as another manifestation of Harmonizing, joins intuition and emotion as a third key element in our ethical selves.

Finally, I would make a more aggressive response to the "we are not good!" position to set alongside the gentler ones. We who cluck or despair or rage at what we see as a pattern of human iniquity are, among other things, Harmonizing with others in an us–them community of virtue, or imagined virtue, that we create by elevating our worthy selves—we who have not lied to dying patients, obeyed Milgram's instructions, married our sisters, eaten our parents, owned slaves, or killed other people—over the unworthy others of our species. It may be that a self-enhancing combination of optimism about ourselves and pessimism about our species is a default position that is inscribed within us even if we consciously reject it. And yet, that particular form of us–them Harmonizing—"good us, bad human race!"—should trouble us. Why are we who judge our species so special, so elevated above the rest of us, exactly?

It is neither ethically appealing, nor likely true, to regard oneself and one's associates as morally superior to humanity as a whole. Instead, I would suggest that it makes sense ethically, as well as being in accord with the facts of the matter, to reach the following conclusion: Yes, I and the people I know are pretty decent, all in all—and so is the species of which we are all examples.

* * *

> In every system of morality, which I have hitherto met with, I have always remarked, that the author proceeds for some time in the ordinary ways of reasoning, and establishes the being of a God, or makes observations concerning human affairs; when all of a sudden I am surprised to find, that instead of the usual copulations of propositions, is, and is not, I meet with no proposition that is not connected with an ought, or an ought not.
> —David Hume, *A Treatise of Human Nature*

Toward an Optimistic, Self-Critical, Self-Improving Ethics

I want to pose two sets of questions connected to Hume's famous quote above, with one set posed in terms of reason and the other posed in terms of ethics. First, some questions from the domain of reason: Suppose one thinks that Harmonizing human nature is indeed good, or reasonably so. What if anything follows from that? Is there a bridge from our understanding of our human nature to our views about how to lead our lives and to collaborate with other people? Or is there instead a gulf, with our understanding of what is on one side—the south rim of the Grand Canyon—and our beliefs as to what ought to be on the other side? Is there a link between Hume's largely optimistic view of human moral sentiments, his critical spirit in denouncing claims to ground morality in reason, and his genial energy in writing, corresponding, and socializing with luminaries—Adam Smith, Rousseau, Diderot—and nonluminaries?[7] And how about the approach to human nature espoused here under the Harmony game framework? Is that framework connected to any "oughts," or is it detached from them?

Turning to ethics: How should we negotiate the complex dance of social ethics in our lives? Does the combination of optimism about people, critical and self-critical spirit, and self-improving energy in Hume constitute a good ethical model for the rest of us to embrace, as part, if not the whole, of ourselves? Or is that ethic flawed? Or is it an ethic that, whatever its value for a limited number of people with the good fortune to have Hume's very high skills and choices, is of limited applicability to the large majority of us who have lesser intellectual and interpersonal talents?

I'll return to the ethical questions. First, I want to address the ones related to reason, using Hume's approach to knowledge—his epistemology—as the starting point. In Hume's framework, there is no bridge of universal reason that will take one from the "is" of science to the "ought" of ethics, but there are a variety of psychological paths that will take one from one emotion-based ethical proposition, such as "human nature is good though imperfect," to another emotion-based ethical proposition, such as "we should be less tribal in our politics."

I think Hume is right about psychological connections between one ethical proposition and another; I believe that there is a connection between optimism about human nature and a self-critical, self-improving spirit. But I first want to suggest that, even assuming Hume is right about the lack of a necessary connection based on reason between scientific truth A and ethical proposition B, human psychology works in such a way that there sometimes is a link. Some scientific truths are pleasant to us, while others are unpleasant, and these different kinds of truths lead us in different directions ethically.

Consider two potential scientific truths about people's ethical nature. The first is our basic contention: Typically, people intuitively treat ambiguous social interactions with other people as Harmony games. The second is the reverse of that proposition: Typically, people intuitively treat ambiguous social interactions with other people as less pleasant games, in which reason and ethics are completely, or partially, opposed.

I hope the discussion in this chapter and the last has gone some way to persuade the reader that the Harmony proposition is truer than the opposing proposition. But regardless of whether that is the case or not, I would suggest that both of the opposing propositions have real psychological connections to opposing ethical propositions.

Our belief in the proposition, "Typically, people intuitively treat ambiguous social interactions as Harmony," psychologically inclines us, though it does not logically compel us, to accept the ethical proposition, "We ought to live in accord with human nature." Correspondingly, our belief in the proposition, "People intuitively treat ambiguous interactions as Disharmony," psychologically inclines us to accept the negative ethical proposition, "We ought not to live in accord with human nature," and to seek out alternative ethical propositions, such as Kant's "We ought to do our moral duty," or Bentham's "We ought to uphold the interests of the whole."

Now, let's turn to connections between one set of emotions and another. The thought here is that an optimistic understanding of human nature such as the Harmony perspective goes along with dual moral emotions: first, a reality-based optimism, and second, a psychologically based self-critical, self-improving spirit toward oneself and groups with which one identifies. Similarly, a pessimistic understanding of humans as not a good lot goes along with parallel dual moral emotions: first, a reality-based pessimism about oneself and one's associates, and second, a psychologically based feeling of being better than other people.

The suggested psychological reason for why an optimistic view of human nature rather than a pessimistic one is a spur to a self-critical, self-improving spirit is a simple one: One of our key moral emotions is competitiveness. We seek to be better than the reference group of all human beings generically. To the extent we believe that our human peers are a fairly sorry lot, we have a low bar to hurdle over to satisfy our competitive goal of being better than they are. But if we believe that human beings as a species are impressive, we have a much more challenging task in lifting ourselves and groups with which we identify above the human norm.

Does the argument I have made contain its own negation within it, as psychological arguments sometimes do, because the psychological link proposed is tenuous and fragile? In other words, we might worry that what applies at one moment, with naive subjects, will no longer apply over time. People will look at their motivations, and in doing so no longer be driven by them.

So, suppose I believe that human nature is good and I then proceed to criticize and improve myself, and groups with which I identify, so as to outdo the human crowd, just as the proposed psychological link suggests. But will I be undermined in my efforts once I realize that I can avoid the hard work of self-criticism and self-improvement by being negative about human nature? Or even if I am not undermined, is there a natural logic of self-interest that will lead most people to adopt negative views of human nature and in so doing avoid hard work?

The claim that self-interest will undermine hard work engendered by a self-critical spirit has a subtler twist. The simple self-interest argument is relatively easily countered. Hard work may be in people's personal self-interest, and even if the benefits largely accrue to others and the hedonic balance is negative for oneself, community norms may well support the self-critical, self-improving meme that in turn supports hard work. But in that case, why does one need the meme? Why not just have the hard work, without the rigmarole of ambivalent moral emotions such as self-critical striving to be better than others?

To the subtler self-interest argument, the right response, I believe, is a variant of the claim about ambiguity that is central to this book. We do not know, most of the time, whether our social reality is Harmony, or some other game. If our own feelings and interests and others' feelings and interests were transparent, everything would be different. But feelings and interests are anything but clear in practice, and given the fundamental ambiguity as to what we and others feel and want, mixed moral emotions, such as our simultaneous competitiveness and shame at our competitiveness, are not a distraction from getting what we want, but constitutive of what we want.

Another potential destroyer of a hard-working, self-critical, self-improving equilibrium is a self-critical, universalist spirit that is simultaneously Choleric and Melancholic. When I reflect on the desire to be better than others that is postulated here to be the cause of my self-critical and self-improving drive, am I seized by a sense by a sense of the ethical wrongfulness of my competitive desire to be better that leads me to abandon my efforts? And if I am not, what about others with more sensitive, or highly developed, Choleric–Melancholic consciences? Might enough of us renounce our competitiveness, and thus destroy the hard work based upon it?

Under the moral emotions approach to human nature advanced here, the possibility of an ethics-based renunciation of effort should not be viewed as contrary to the fundamental logic of human nature. But neither, I think, is it a possibility to be unduly worried about, or for that matter unduly hopeful about. Melancholy shame and self-punishing Choleric guilt are powerful emotions indeed within us—but social communities that are not so overcome by these emotions as to abandon competitive striving will have an advantage, all else equal, over communities who are so powerfully swayed by shame and guilt that they abandon struggle.

Now, though, we need to enter the domain of normative ethics. *Should* I be seized by a sense of the ethical wrongfulness of my competitive desire to be better than others? I believe the right answer is no. It seems to me important to avoid succumbing to egoistic bias in regarding myself, or groups with which I identify, as ethically better than other people or groups, and further important to feel that I am, and groups with which I identify are, in moral parity with rather than morally above most other people and most other groups.

But it is not good, I would suggest, to believe that striving to be ethically better than others is bad. It is true, and important, that in most social settings, announcing a goal of being better ethically than another individual or the group as whole is inconsistent with effective Harmonizing with that individual or group. But to experience competitive urges and emotions that lead one to act better is fine, even if these urges need to be kept quiet, and indeed apologized for, under valid protocols of Harmony.

To restate the point in more general terms: One's competitive striving to be better than others, or than one's past self, in being forbearing, temperate, just, and otherwise virtuous is an instance of seeking the good for something other than its own sake. But that striving is good, under the account proposed here, so long as the game one is constructing with others through moral striving is a consensual Harmony game.

When we challenge people to improve, we need, in the account of human emotion and reason suggested here, to strive for a Harmony game in which everyone in the group we are addressing has shared interests. It may well be the case that only some in the group will in fact share common interests and be inspired, but the call for moral betterment, if it is to be morally compelling, needs to be constructed in the form of a Harmony game in which everyone can participate, and no one is excluded.

In the universal form of a Harmony game that is represented by writing, with an audience that could include anyone, the scope of the community that needs to be offered a Harmony game if one's appeal is not to be self-undermining becomes universal.[8] One is appealing to a particular group, such as social psychologists or business ethics professors and students, but also to human beings generically. Under current conditions, that universal game frequently subsumes the more limited Harmony game of speech; spoken words are ever more often spoken with the awareness that they may be heard or read by anyone.

We can now return to the questions about human nature and ethics posed at the beginning of this section. I believe that optimism about people, a self-critical and self-improving spirit, and effectiveness do in fact constitute a valid ethical model for all of us, regardless of our station. If I do not believe that—if I try to play one kind of Harmony game by saying that I will not try to persuade you of the ethical merits of optimism about people, or of the self-critical spirit that I believe is related to that belief—I am not playing another important Harmony game, the game of reason, correctly.

Since Hume's emotion-based approach to ethics seems to me mostly right, I am constrained to also believe that my ethical belief in certain qualities as worthy of universal adoption is itself a sentiment, and a sentiment that is influenced by my own character and my circumstances. That acknowledgment provokes further questions. Can it really be true that any given ethic, such as the Phlegmatic ethic of business, is an ethic applicable to all of us, Western and non-Western, liberal and conservative, modern and ancient alike?

If Harmonizing is different in Lagos and Orlando, in Hume's Edinburgh and in my New Jersey, are there not different, and possibly equally appropriate, ethics corresponding to different places, peoples, and eras? We return to that question later, in the final chapter. To anticipate the discussion there: I believe all of us should indeed have the spirit I call business ethics inside us as part of our moral emotions. But I believe we all should also have other spirits, other paths to Harmony, within us as well.

* * *

> Virtue is concerned with passions and actions, in which excess is a form of failure, and so is defect, while the intermediate is praised and is a form of success; and being praised and being successful are both characteristics of virtue.
> —Aristotle, *The Nicomachean Ethics*

> [The people say:] Suppose now that there were two such magic rings, and the just put on one of them and the unjust the other; no man can be imagined to be of such an iron nature that he would stand fast in justice. No man would keep his hands off what was not his own when he could safely take what he liked out of the market, or go into houses and lie with anyone at his pleasure, or kill or release from prison whom he would, and in all respects be like a god among men.
> —Glaucon to Socrates, *The Republic*, Book II

Living in Accord with Healthy Human Nature

As a young man, I disagreed with Aristotle on ethics before I ever read him. I was a loyal devotee of Bertrand Russell and his *History of Western Philosophy*, and Russell's disdain for Aristotle's ethics—"Aristotle's opinions on moral questions were always such as were conventional in his day"—was also mine. As a late middle-aged man, there now seems to me a great deal of truth in Aristotle's linking of ethics and worldly success, as well as in his practical, socially oriented view of ethics. Aristotle's definition of virtue, with only slight modifications, works as a statement of the Harmony perspective of this book. Human beings, in the view advanced here, are evolved so as to agree to an extremely high degree in our respect for virtue, to a very high degree on what brings us closer to virtue, and to a high, though imperfect degree in aligning intuitively to practice it. Given these aspects of human nature, living our lives in accord with it is a worthy ideal for us to be guided by.

The ambivalent, Choleric, radical, literary, dialogical sensibility of Plato appealed to me greatly as a young man, for the same reasons I favored the similar temperamental spirits of Nietzsche and Dostoevsky. It still does. Glaucon's story of the ring poses the question of whether living in accord with ordinary human nature is in fact a good thing. By having Socrates praise Glaucon for his acuity and move on to another subject, Plato has allowed his readers over the last two thousand years to give their own answers to that question. In this chapter and the last one, I have given my own, affirmative response. I conclude my discussion of Harmonizing human nature in Socratic, Glauconian terms.

First, I would suggest that the Athenians as Glaucon describes them exemplify a general negative bias in judging human nature that arises from the greater salience of other people's negative intentions and behavior than of their positive intentions and behavior. While our intuitions lead us to collaborate with other people to construe our shared social reality in terms of Harmony games, the Choleric, righteous, self-righteous part of our moral sense, as well as a good part of our Melancholy side, focuses on the negative aspects of what other people, and sometimes, we ourselves, feel and do. Even if ninety-nine of a hundred people would act more like J. R. R. Tolkien's decent, though imperfect, hobbit Frodo than like his evil wizard Sauron if they came into possession of an evil ring of power by trying to destroy the ring, faced with such a sample of one hundred, one Sauron would compel our attention more than any of the ninety-nine Frodos. That would lead many of us to judge our nature as humans in the jaundiced way the Athenians did.

Second, I would suggest that human beings in Glaucon's times and ours alike are wedded to the drama of small and large social traumas. We focus on the negative aspects of the us–them moral proclivities of our species. In doing so, we tend to overlook our remarkable capability for treating ambiguous social situations with other human beings in terms of Harmony games in which we are aligned with them. We are a species whose remarkable rise owes a great deal to our intuitive ability to Harmonize.

I believe that we underestimate people morally by setting implausibly high expectations of what it means to be good. It is good, I would suggest, to adopt a modest ethical goal of living in accord with healthy human nature. We nowadays are similar to Glaucon's Athenians in holding radically high moral standards that we see ourselves as falling short of. We, like they, doubt that people are truly virtuous, as opposed to being interested in getting along and getting ahead through seeming to be virtuous. In holding ourselves to an elevated, extremely hard-to-meet standard of seeking the good for its own sake, we, like the Athenians, condemn ourselves. But isn't the social alignment with other human beings that we seek on an everyday, wired-into-our-nature basis, itself, good, even if it is not the good in itself?

I contend that it is.

To make a different, nonargumentative, face to face across the millennia response to Glaucon's Athenians: We are different in some ways from you. But we, and our ethics, are not morally superior to you, and to your ethics. Your stories of shepherds and magic rings are also ours, and our stories of experimenters in lab coats and accountants in corporate frauds would also be yours, if only we who live now could ever walk among you. We may look down on you if we wish—how could you condone slavery, and certain other forms of radical social inequality, with the equanimity you did? We have done well, and achieved genuine insights and advancements, with Melancholy, Choleric, and Phlegmatic reason. But looking down from the Melancholy heights of obedience to moral principle, the Choleric heights of righteousness and self-righteousness, and the Phlegmatic heights of calculating reason is not all there is to ethical elevation. There is also looking down from the Sanguine heights of ethical affirmation—and there, you are in the highlands and we are in the lowlands (Figure 2.1).

The Eight Kinds of Harmony Game

Temperament	System 1—Intuition, emotion	System 2—Reasoning
Sanguine	Strong	**Weak**
Phlegmatic	Strong	Strong
Choleric	Strong	Strong
Melancholy	Strong	Strong

People now, as in the past, are very skilled at System 1, intuitive, emotionally based Harmonizing in Sanguine, Phlegmatic, Choleric, and Melancholy moods. We are also very skilled in employing System 2 reason on behalf of Phlegmatic, Choleric, and Melancholy Harmony. Where we lag, perhaps especially in modern times, is in employing reason to create and sustain Sanguine Harmony.

Figure 2.1 The Weakest Link

Summary

Human beings constantly create Harmony Games of different kinds with other human beings. We Harmonize in a competitive, Choleric, aggressive spirit in games that are called zero-sum, but in which the players are united by positive-sum, shared desires to win, and to act in accord with the spirit of the game. We Harmonize in a Melancholy humor in games, sometimes ritualized and sometimes spontaneous, in which we are united in reactive, negative feelings, such as sadness, boredom, discouragement, isolation, anxiety, shame, and humiliation. We Harmonize in a Phlegmatic, calm, practical, humor in games we play with business colleagues, and with unemotional, inert matter. Finally, we Harmonize in the happy, flourishing, changeable, positive, active Sanguine spirit that, more than any other, epitomizes Harmony.

In addition to our fluid, socially smooth, intuitive, System 1 Harmonizing, we Harmonize, sometimes through writing and reading, in slower, reasoned, more

(continued)

> *Summary* (continued)
>
> socially awkward, System 2 forms. The temperament that is hardest to blend with System 2 Harmony is the Sanguine. Our modern intellectual culture does much better with Melancholy, Choleric, and, especially, Phlegmatic System 2 Harmony than it does with the Sanguine form. In both their System 1 and System 2 forms, Harmony games are ethically imperfect. They accompany, and may facilitate, evil. Harmony is nonetheless good, as is our drive to create it.

> *Exercises*
>
> 1. Pick a work of literature by a highly regarded author and analyze a few pages of it. Do all four temperamental forms of Harmony appear? What about System 1 and System 2 forms of Harmony?
> 2. (a) Pick a work of popular fiction and analyze a few pages of it. Do all four temperamental forms of Harmony appear? What about System 1 and System 2 forms of Harmony? (b) Draw, or search online for, an illustration of Harmony in multiple forms.
> 3. (a) Pick a philosophical work on ethics—scriptural, classical, or modern—and analyze a few pages of it. Do all four temperamental forms of Harmony appear? What about System 1 and System 2 forms of Harmony? (b) Pick an academic management book or article and analyze a few pages of it. Do all four temperamental forms of Harmony appear? What about System 1 and System 2 forms of Harmony? (c) Do the quiz in the endnotes, and discussed in Chapter Five, on whether people's reactions to defecting in a Prisoner's Dilemma are more affected by how well they do, or whether they are aligned with the other player.
> 4. (a) Pick a popular management book or article that describes the author's or other managers' experiences and analyze a few pages of it. Do all four temperamental forms of Harmony appear? What about System 1 and System 2 forms of Harmony? (b) Watch a video of a management lecture, or watch your teacher. Do all four temperamental forms of Harmony appear in the interactions of the lecturer (or teacher) with the audience (or students)? What about System 1 and System 2 Harmony?

CHAPTER 3

Opening the Door to the Sanguine

Standard mathematical game theory, as expressed in matrices, game trees, equations, and other forms, is, in my view, both beautiful and highly valuable. Further, the standard game-theoretic stories of prisoners and so on are highly interesting, and illuminating. But it is a different version of game theory, a critical, philosophical, moral emotions version that does not assume the calculating egoism that is the common currency of the standard tales of prisoners, teenage daredevils, and the like, though not of mainstream game theory as a whole, that this book is devoted to advancing. It is important not to assimilate the philosophical approach to game theory advanced here to the standard mathematical approach, and to the standard stories.

While asserting separation, I want to note the deep indebtedness of the critical approach taken here to mainstream game theory.[1] In particular, the Four Temperaments approach to game theory is indebted to the applied version of mainstream theory I learned from Professor Schelling, who stressed to us that human motivations were varied and that nonegoistic motivations, and the commitments that such motivations made possible, could often help players both to succeed and to achieve collectively desirable outcomes. Though I would not have expressed it that way at the time, I revered him then as a man whose ideas—expressed in his 1960 book, *The Strategy of Conflict*—supplied a language that allowed American and Russian leaders to see themselves as partners with certain shared interests, and I revere him now as someone with a normative vision that I believe contributed to the calming of the Cold War, and to our all being here today. My hopes for the good that may come for individuals, organizations, and society from the version of game theory advanced here are considerably more modest than that, but they are derived from the fundamental faith in a linkage between logical reasoning and changing the world for the better that I gained from his analysis of nuclear deterrence, racial segregation, and other topics.

In all the stories that will be told in this chapter, the broad aim, if not necessarily the immediate one, will be to open our minds to reasoned Sanguine alternatives to the Choleric, Melancholy, and Phlegmatic forms of reason to which we are accustomed. I have a main priority and two preliminary ones. The main priority is to explore what happens if we flip the assumption of player egoism in standard game-theoretic stories like the Prisoner's Dilemma in favor of an assumption that players have pro-social motivations. When we do that, I suggest, something very interesting happens. A switch is flipped. The pro-social, yielding junior and senior workers in Deference, our first flipped story, and the protagonists in the other flipped stories that will be told in this chapter have real problems in solving difficult games, just as the egoists do. But one is able to view their travails in difficult games, and the possibilities for addressing them, in a charitable fashion, in contrast to the harsher mode in which one looks at the woes associated with egoism. Further, in the one easy game, Harmony, flipping helps one become empowered to move, at times, from pragmatic calm to Sanguine joy in contemplation and creation.

To arrive at our destination in this central chapter of the book, we will first prepare for the journey. Two preparatory steps are involved. First, I briefly explain the "Knobe effect," which helps us understand how human moral intuition regards consequences as intended or not, and to which I return in Chapter Five. Next, I briefly explain the games of interest. In the first two chapters, we focused on Harmony. Here, all the games are in play. I introduce a fourfold classification of games, indebted to my teacher Thomas Schelling, under which they can be understood in terms of two logics—or ethics—Dominance and Highest Joint Value, which are sometimes in accord, and sometimes in opposition.

The Blame Game

Several years ago, experimental philosopher Joshua Knobe asked a number of random people walking in Central Park a question about whether a profit-seeking CEO intends to harm the environment by adopting a new product that will increase profits, and also damage the environment. He asked a different random group about whether a profit-seeking CEO intends to help the environment by adopting a new product that will increase profits, and also improve the environment. Knobe found a striking asymmetry in the results. A large majority of the "harm" group said the CEO intends the harmful consequences, while a large majority of the "help" group said the CEO did not intend the helpful consequences.[2]

With my Rutgers Business School undergraduate and MBA students, I've been drawing on Knobe's intriguing finding to explore how people respond to the actions of principle-following actors. The bottom line is very simple: My students—who are, I believe, representative of human moral intuition, like Knobe's Manhattan strollers—do not believe that a principle-following actor intends the negative consequences of his or her action. My key question for them is the following: Does a "we can't harm the environment!" CEO intend to harm the company by refusing to adopt a new product that will increase profits, because it will also damage the environment? No, is the clear answer.

The valuable contentious, Choleric spirit within us will have concerns and arguments to raise about Knobe's survey, and my extension of it. I defer those until Chapter Five. Here, the purpose is to suggest the following proposition as an introduction to our inquiry: We are so made as to view the consequences of egoistic calculation with an eye toward blame, not toward praise or joy. By contrast, we view the consequences of pro-sociality in a different, much more benevolent manner.

A Classification of Social Games: Harmony, Imperfect Harmony, Partial Disharmony, and Disharmony

First, an acknowledgment: In an article I'm sure I read for Professor Schelling's seminar, though I don't remember doing so, he classified games according to whether the best overall outcome for the players requires none, all, or some of the players to do something other than what is in their individual, immediate interest.[3] The fourfold classification of games presented in this section is indebted to his more elaborate classification.[4] (Interested readers are invited to refer to Appendix A.)

Now, the definitions. The classification of social games in this book is based on two fundamental concepts: Highest Joint Value, derived from Schelling's classification, and Dominance, a broad, widely applied, concept in game theory.[5] If one follows the logic of Highest Joint Value, one acts so as to make possible the best outcome for the whole, assuming the other player, or other players, do likewise. If one follows the logic of Dominance, one plays Dominant strategies—that is, strategies that are better given one's feelings, interests, and values (which may be altruistic, egoistic, or anything else), regardless of what the other player does.

In Sanguine, benevolent Harmony Games, the Highest Joint Value and Dominance principles are in accord. What the players want—which may be the result of their generous feelings, their norm-following feelings, their competitive feelings, or other social feelings, rather than the result of their egoistic feelings—is aligned. The counsel of the heart and the head are the same. If one reflected, one would say, "I care how she acts—but it doesn't change what I do." There are many Harmony games, as we have discussed in the last two chapters. We will introduce new, flipped ones in this chapter, such as I Get It!, Gratitude, and Love. But all of them share a benevolent logic.

In Phlegmatic, Melancholy, and Choleric Games, on the other hand, Highest Joint Value (HJV) and Dominance are not perfectly aligned. In Phlegmatic, or Imperfect Harmony Games, there is a considerable degree of concordance between the logics of HJV and Dominance, but also some degree of tension. In Imperfect Harmony games, one's best response as either an egoist or an altruist to the other player playing HJV is to do likewise. In that sense, there is a major degree of Harmony. But Harmony is only partial. If the other player does not play HJV, one's best response in an Imperfect Harmony Game is also to deviate from HJV. One could view that deviation in egoistic terms—one does not trust the other player—or in altruistic terms—one does not want to embarrass or shame the other player. But in either case, there is dissonance as well as consonance in Imperfect Harmony Games.

Sensitive Boy, in which a prehistoric mother and a father are considering what to do in rearing a son who does not want to play hunting games with the other boys, and Follow the Rule, in which two prehistoric mothers are out separately with their children, and are deciding whether to make their children pick only fruit or also allow them to pick berries, are two flipped Imperfect Harmony stories that will be told later in this chapter. The mothers and the father do not want, we will assume, to show up or embarrass their counterparts. Worthy as that pro-social goal is, it can go along with difficulties in their reaching the best outcome, as we shall see, just as is the case with more selfishly motivated actors.

In Melancholy, or Partial Disharmony, Games, we move from a predominance of Harmony to a predominance of Disharmony. Specifically, in this type of game one's best response to the other player playing HJV is for you to deviate from HJV. One can view that deviation in altruistic terms—one sacrifices one's own needs and the greater good of the whole in order to help the other—as well as in egoistic, or competitive, terms—one takes advantage of a cooperative other by bullying, or slacking. Either way, there is a major obstacle to the players both playing HJV in a Partial Disharmony game. On the other hand, such games have an element of Harmony. If the other player deviates from HJV, you have an incentive to adhere to it.

Where Harmony is fully lost is in Choleric, or Disharmony, Games. In these games, Dominance reasoning and the logic of HJV are at war. In Disharmony, the best outcome for the players combined and/or the broader unit of which they are a part entails their playing what game theorists call a "dominated strategy," one that does *worse* no matter how the other acts. Players in a Disharmony game are thus faced with a nasty problem as to whether to follow HJV reason or Dominance reason. With Disharmony, we complete the spectrum. We have moved from our starting point in easy Harmony on to moderately difficult Imperfect Harmony, then to difficult Partial Disharmony, and finally to very difficult Disharmony.

In our flipped stories, as it happens, Partial Disharmony and pure Disharmony merge. Managerialism, in which a board and a CEO are locked into busywork when both could do better with shared leadership, illustrates both kinds of game. So does Deference, in which the senior and junior worker would both do better if only they could assert together.

* * *

> *To be filled with your masculine power—the Yang,*
>   *follow your feminine nature—the Yin.*
> *Be a valley under Heaven.*
> *Be a valley under Heaven and your potency*
>   *will not fade away.*
> *You will become like a little child again.*
> *To be filled with the light within you—the Yang,*
>   *follow the dark within you—the Yin.*
> *Be a model under Heaven.*
> *Be a model under Heaven and your potency*
>   *will not fade away.*
> *You will return to the Infinite.*
> *To be filled with honor—the Yang,*
>   *follow humiliation—the Yin.*
> *Be empty as a valley under Heaven.*
> *Be empty as a valley under Heaven and your potency*
>   *will not fade away.*
> *You will return to the uncarved block.*
> *The uncarved block is cut up*
>   *and made into useful things.*
> *Wise souls are carved up*
>   *to make into leaders.*
> *Just so, a great carving*
>   *is done without cutting.*
>   —Lao Zi, *The Dao-de-jing*, Chapter 28

[S]o frivolous is he that, though full of a thousand reasons for weariness, the least thing, such as playing billiards or hitting a ball, is sufficient to amuse him. But will you say what object has he in all this? The pleasure of bragging tomorrow among his friends that he has played better than another. So others sweat in their own rooms to show to the learned that they have solved a problem in algebra, which no one had hitherto been able to solve . . . In a word, man knows that he is wretched. He is therefore wretched, because he is so; but he is really great because he knows it.

—Blaise Pascal, *Pensees* (1670)

In this section, I begin the project of flipping the paradigmatic game-theoretic stories from tales of the travails of egoistic and competitive reason into very different tales, beginning with the uber-story of the game-theoretic canon, the Prisoner's Dilemma. As noted, the aim is not to debunk the classical stories. They do a very fine job, I believe, in anatomizing certain parts of our nature. Part of us—a bigger part in some of us, but a real part for all of us—is calculating, self-concerned, and proud, or, in the language of the temperaments, Phlegmatic and Choleric. That egoistic, calculating, aggressive, competitive part of us is nicely described by Pascal in his self-critical description of the mathematician. Classical game theory, with its stories of prisoners, daredevils, hunters, and battling spouses, does very nicely indeed as a portrait, an often Melancholy one, of how this part of ourselves encounters difficulties in solving a range of social games with other people.

The aim in what follows is to anatomize a very different side of our nature, evoked by Lao Zi, as well by Pascal in his second quote on human shame and its greatness. That ashamed, sociable, sympathetic, and empathetic side of ourselves—to follow Lao Zi, our Yin, or classically feminine, side—has its own dilemmas of reason that parallel, though they are not the same as, those of calculating, egoistic reason. Through understanding Harmony, Disharmony, and the games in-between in a different way, one that flips the standard stories to create new stories, we can, I suggest, paint a different portrait of how we fail and succeed at social games. In creating these alternative pictures of ourselves, we can, I suggest, open the door to Sanguine forms of feeling and reason.

The most powerful standard game-theoretic story, and a good contender for the most powerful new one-paragraph story of the last hundred years, is the Prisoner's Dilemma. Recall the version Steven Pinker tells, the Pacifist's Dilemma, which he also calls the Tragedy of Violence: If only we—two bands of early humans—could refrain from aggression against the other, we would both be better off. But no matter what the other band does, we are better off being aggressive. If they are not aggressive, we gain resources by aggression against them. If they are aggressive, we and they both lose when we fight back, but we lose less than if we let them prevail over us. If only we could both avoid aggression, it would be wonderful. But how is stable peace possible, given that aggression is always a best response?[6]

I noted in my first go-round with Pinker's Tragedy that our long-ago ancestors, like us, had ambiguous payoffs, and that they, like us, could have interpreted, and, I suggested, mostly did interpret, ambiguous reality in Harmony terms rather than in Tragedy terms. My point here is very different: The Tragedy/Dilemma misses out on the Sanguine, even as it powerfully evokes the other three quadrants. It is conducive alike to Shame ("Thou shalt not defect!"), to Choler ("Punish the defectors!"), and to a bemused, detached Phlegmatic spirit that observes the split between individual and group advantage without judgment. What is lacking in the standard story of the Dilemma is not support for altruism, or for the related social emotions of sympathy and empathy—one can say the story is in fact about the need for these qualities—but a happy spirit. The world of the standard Dilemma—or the world of Disharmony, to use my term—is an anxious, sometimes sad, sometimes angry one in which we reason out how to apply internal and external sanctions to save us from ourselves.

If it is good, as I believe it is, to combine System 1 ethical passions with System 2 ethical reasoning, is there a way that we can keep the rational spirit of the Dilemma, but open it up to the Sanguine element that is missing in the standard account? I believe there is. Let's start on that path by considering what is arguably the single most central relationship in business, the one between managers and employees.

The Manager and the Employee: Deference

Let's first consider how the standard Disharmony story can be retold for an egoistic manager and employee. In this version—we can call it Slacking—there is a project on which both are working. If only they could both commit fully to the project, it would be better for them and their organization. But it is better for both of them as individuals to slack off to some degree, no matter what the other does. If the other fully commits, you can get credit without doing much work. If the other slacks, you do better by slacking as well, rather than by being the sucker who does the bulk of the work. We thus have Disharmony: The logic of Dominance tells you to slack, even as the logic of HJV tells you to commit.

There is a long-established body of Melancholy, Choleric, and Phlegmatic managerial reason—ranging from the time of the pyramids to Frederick Taylor's Scientific Management in the early twentieth century[7] to the present—that takes the slacking story very seriously and tries to deal with it through internal and external incentive mechanisms. On the other side, advocates of Human Relations and similar approaches to management[8] have tended to doubt the veracity of Slacking—"People actually want to commit!"—or to worry about its consequences—"People will live down to what we expect of them." In turn, advocates of the standard story question its critics—"We live in the real world, not in a world in which wishes are ponies." A different, and I think more promising, strategy is for all of us to give the standard slacking story its due as capturing one very important part of who we are, to avoid Choleric complaining about the story, or about its critics, and to consider an alternative story that flips Slacking.

We can call the flipped story Deference. In it, we have a manager and an employee who are motivated not by egoism and competitiveness but by sympathy, empathy, and shame. Imagine a senior manager who has reached the highest level he or she expects to reach, and who cares about the wellbeing and the development of a junior employee. Imagine a junior who respects the judgment and character of the senior and wants nothing more than to do what the senior wants. They, and all of us—for all of us have sympathy and shame as part of our makeup—face their own version of Disharmony. If only you could both self-assert and commit to your individual advancement, you and your organization could both do better. But no matter what the other does, you are better off Deferring. If the other Asserts, you are happier following than Asserting yourself. If the other Defers, you do not want to make the other look bad and hog the limelight by Asserting. So you Defer, regardless. And that is too bad. Thus, the alternative version of Disharmony: Dominance tells us to Defer, and in so doing clashes with HJV, which tells us to Assert.

Now, the key step in the flipping process. The observable behavior in Slacking and in Deference is, we may assume, the same. So are the outcomes. If the players in both games follow the ethic of Dominance, they are less happy than they could be, and their firm is less productive than it could be. But the psychic qualities that produce the outcomes are very different. In Slacker, we assume egoism. In Deference, we assume altruism. We wind up with the same actions and the same results, but for very different reasons.

What follows? First, the flipped story allows us to see that the problem of Disharmony is not at all an artifact of egoism. Disharmony is just as present in Deference, with its altruistic manager and employee, as it is in Slacker, with its egoistic manager and employee. Altruism, sympathy, empathy, shame, and guilt, valuable as they all are, do not lift us above Disharmony. Instead, they create their own versions of Disharmony, which may be identical in their outer manifestations, though not in their inner motivations, to the egoistic and competitive versions.

Second, and crucially, the flipped story opens the door to Sanguine reason in a way that the standard story does not. In Deference, one has a problem for System 2 to tackle, just as one does in Slacking. Just as one can propose mechanisms to deal with individually and organizationally deleterious Slacking, one can propose mechanisms to deal with individually and organizationally deleterious Deference. But in Deference, unlike in Slacker, the proposed System 2 ideas to improve the situation start from a Sanguine place that affirms the motivations of the deferential manager and the deferential employee.

It is a mistake, I believe, for managers to neglect the Melancholy and Choleric forms of reason distinctively promoted by Slacker, and more broadly by the traditional Prisoner's Dilemma/Disharmony story, in favor of the Sanguine form of reason distinctively promoted by Deference, and by other flipped stories. Much as one can and should abide by norms of Harmony that make an appeal to the Sanguine or the Phlegmatic appropriate, and frown on direct managerial invocations of shame, guilt, and retribution, dark-side human emotions are not at all to be disdained in either System 2 or System 1 management. But there is, I believe, a very great value in our being able to appreciate Deference and its peers alongside Slacker and its peers, and to see Sanguine reason not as a leap of faith, but as a branch of System 2 reason alongside its Phlegmatic, Choleric, and Melancholy peers.

Just as one can spin out many standard Disharmony/Dilemma stories, so too with flipped Disharmony stories that start from an optimistic place about the players. For example, consider a manager and an employee who embody what one believes is an ideal balance among temperamental qualities, between System 1 and System 2, and between technical and human aspects of their work. Now the Disharmony story works as follows: If only the two of you could both be extreme (with, say, one of you concentrating on coding, and the other on meeting clients; or in a temperamental version, with one of you tilted toward Choler and the other toward Melancholy rather than balanced, both of you, and your organization, would do better. But no matter what the other does, you are better off maintaining balance. You are in Disharmony, much as both of you are highly estimable, balanced human beings. You need to apply Sanguine reason to figure out how to be more extreme in the situation, or class of situations, in which that is what is called for.

The Virtues and Game Theory

Let us now consider an objection, or a concern, that can be expressed as follows:

> What you are saying boils down to a claim that the worthy parts of ourselves, or our worthy selves as a whole, are prone to their own problems, just as the less worthy parts, and less worthy selves, are. I understand the logic, but I have

three questions. First, two well-balanced people are not nearly as likely to be in Disharmony as two people with less well-balanced characters. Is that not so? Second, you have told us before that people in business and elsewhere typically do not know what their payoffs are, and hence converge on playing the most pleasant game, Harmony. So why should we care about Disharmony, or any game other than Harmony? Third and finally, I am not sure how your concept of Sanguine reason in the end differs from the time-honored concept of cultivating the virtues. If I recall, you were concerned that virtue ethics, compared to its utilitarian and deontological rivals, is conducive to bromides rather than to the subtler forms of System 2 reason. But is there a technology of Sanguine reason that you are proposing? If not, fine. But in that case, what you are offering is simply another way of grounding the standard virtue ethics case for balance, with the implementation issues unaddressed. Is that not so?

I address the questions/concerns from last to first, beginning with the question of whether the approach proposed here in the end boils down in practical terms to the traditional virtue ethics concern with the cultivation of character. To that inquiry, my initial response is that, if it were so, that would be fine. To understand the approach advocated here as another way of grounding the virtues is different from what I have in mind, but is not contrary. My second response is that the Sanguine reason, Four Temperaments approach that is advanced here often, though not invariably, does in fact suggest practical answers to the "what should I do?" questions that are the weak spot of virtue ethics, compared to its rivals. I suggested in the last chapter that one can transition from the realization that one has been unduly locked into a certain humor—Choler, say—to playing Harmony games involving different humors. The thought here is that flipping standard models of difficult situations can help one to do a better job in playing games that involve Sanguine reason and feeling.

A brief example of the application of Sanguine reason to a practical situation, drawn from personal experience, slightly modified: I receive an email from my supervisor one evening asking me to have a complex memo that I have not yet started to be completed and circulated the next day. I have lots of other things to do, such as writing my daily quota of words for my book and preparing for a school board meeting at which there will likely be a stressful discussion of a lawsuit. What am I to do? Carry out a cost–benefit analysis? Decide if there is a fundamental moral duty to comply with or not? It occurs to me that the situation can be understood as a case of my boss solving Deference in a micro context. She has Asserted. I can thank her, stop writing my words early, stop worrying about the meeting, and get down to my own form of value-enhancing Assertion by writing the memo in a way that reflects my voice and ideas.

The critic's second concern is that a focus on understanding and solving Disharmony, or other difficult games, clashes with the premise that Harmony is the governing game. Here, the response is again twofold. First, as previously noted, when Pinker's Tragedy was addressed for the first time, we need to have the resources within us to solve difficult games, even though we cannot be sure, as a rule, of the identity of the game we are playing. If we cannot deal with difficult games, Harmony is a dream, rather than realistic. We need to have a part of us that plays the hard games, even as Nature plays those games with us. Second, we can understand Sanguine reason as a tool for the conversion of difficult games into Harmony. Yes, we may be in the Disharmonic state described by Slacking or Deference games (or, on the other side of the excess–deficiency spectrum, by Overwork or Egomania games). But to see Disharmony in the light of Sanguine, as opposed to Melancholy or Choleric, reason is also to see the Harmonic possibility of aligning the basically good natures of oneself and the other.

Finally, I strongly concur with the critic's point that worthy aspects of people's characters, or characters that overall are good, generate less Disharmony than less pleasant aspects of ourselves, and less pleasant characters. But I do not agree that Disharmony games that take an optimistic perspective on the players, such as Deference, are less important than games that take a pessimistic perspective, such as Slacker. We may hope—and our hope may be a reasoned one—that overall, and most of the time in particular cases, the balance of self-concern, sympathy, pride, shame, calm, and other qualities in us is a healthy one. We can reasonably see Deference, and other dilemmas of Sanguine reason, as being as, or more, important in our daily lives at work, and elsewhere, than the dilemmas of Melancholy and Choleric reason. One important reason for adopting that perspective, I would suggest, is that the players can openly discuss among themselves a game like Deference in an equable, Harmonizing spirit. By contrast, standard Dilemma/Disharmony stories are extremely difficult, often well-nigh impossible, to air out openly, given the valid Harmony convention against treating people with whom we are in relationship as greedy, lazy, angry, and so on.

A brief example to clarify the preceding claim: Suppose you are a manager who is trying to reform a work culture in which managers and employees are often less than fully committed. As part of a fused System 2–System 1 approach to improving the situation, it may well be worth seeing if you can bring people together, and then move them forward together, by discussing the issue you face in terms of Deference, or another Disharmony game in which the players have good characters. If you discuss it in terms of Slacking, what you have to say will be understood in terms of the blame-filled operations of Choleric and Melancholy reason, and will engender defensive resentment. If you instead discuss the issue in terms of Deference (or other Sanguine terms), you have a chance for success that you lack when you employ a traditional Disharmony/Dilemma story like Slacking.

The Employee and the Company: Agreeableness

I now want to turn to applying flipping to another relationship critical to business ethics.

In this game, one player is an employee, managerial or otherwise; in a multiplayer game, we can instead think of groups of employees, managerial or otherwise. The other player is the company, personified in the form of the multiple individuals, such as officers and board members, who set company policy and act on its behalf. As a Disharmony story grounded in Choleric and Melancholy reason, I suggest the following: Both the company and the employee could do better for themselves, and the overarching whole of which they are a part, if only they could make promises to high commitment and live up to those promises. The problem is that no matter whether the other side keeps its implicit promises or not, one is better off not keeping one's own. The logic of Dominance thus pushes everyone—for example, a company violating implicit promises to its senior employees by firing them, or a manager or employee failing to live up their implicit promises to be "all-in" rather than detached—to finagle or, more bluntly, to lie and cheat. Thus, the Dishonesty version of Disharmony.

Among mainstream finance scholars, Michael Jensen stands out for his delineation of both sides of the Dishonesty Dilemma. His early, highly influential advocacy of stock options can be understood as a proposal for a reasoned System 2 solution to Dishonesty: Managerial and employee failures to live up to their moral duty to their employers can be mitigated through aligning individuals' Phlegmatic calculations with a broader interest. More recently, he has focused on the problem of lying in budgeting, and yet more recently, in a body of work that is highly relevant for our purposes here, he has turned to the problem of how people in organizations, and by extension organizations themselves, can make effective promises that are upheld reputationally, as well as through conscience—"I'd better keep my promise, and if I don't, I've *got to* apologize!"—in a way that avoids the pecuniary and Harmony costs associated with formal legal contracts.[9]

Like other standard versions of Disharmony, Jensen's Dishonesty story resonates in terms of Choleric and Melancholy reason, and is lacking on the Sanguine side. As a flipped version, I would suggest that we focus on how both the employee on one side, and the firm through its representatives on the other, can be understood as worthy people, or entities, that are motivated not by Dishonesty, but by Agreeableness. Living as one should by canons of Harmony, one avoids unpleasantness. When oneself, or the other, makes commitments, implicit or explicit, that oneself or the other does not fulfill, you are both in an unpleasant situation. Better to let it go by not hassling yourself or the other, the logic of Dominance counsels—you will only disturb the smooth surface of agreeability, no matter what the other does—even as the logic of HJV counsels an open, or Direct, approach. Thus, the Agreeableness version of Disharmony. The outcome and the behaviors are the same as in Dishonesty, but the explanation of what is going on is quite different.

Jensen's late-career effort to create mechanisms for effective commitments can, and I believe should, be understood as one avenue, among many possibilities, for applying System 2 Sanguine reason to help both the employee and the firm (i.e., its representatives) to attain directness that is typically unpleasant, but that is situationally valuable. Some applications of Sanguine reason can be tailored to particular groups and situations, such as advice to female managers and employees on Directness in negotiating on one's behalf,[10] and advice to companies on directness in dealing with sensitive matters of balancing work and family commitments. Other applications, such as Jensen's proposal for apologizing for broken commitments and clarifying new commitments, or gender-neutral, perspective-neutral advice to both employees and firms on directness in negotiating, can have a broader scope.

I close this section with two notes on the Agreeableness version of Disharmony. The first is that the Agreeableness Dilemma between the employee and the firm is similar to the Deference Dilemma between the employee and the manager. In both cases, the message is parallel: Sympathetic qualities and characters generate their own Dilemmas, which can be addressed through Sanguine reason. The second is a more limited point about business ethics. Over the years, Jensen's sharp-edged advocacy for shareholder value as the single criterion that managers should maximize[11] has made him a debate partner in a Choleric Harmony game with mainstream business ethicists. Fair enough—but for those of us who find the Sanguine underrepresented and undervalued in the halls of reason, it would be very fine if, in this late stage of his highly productive, highly provocative career, Jensen and business ethicists started playing a new Sanguine Harmony game together, with his ideas about promise-making and promise-keeping as key components of that game.

Business and Society: Rationality

Let us now turn to applying the flipping approach to a third critical relationship for business ethics, that between firms and societies. In the course of telling that story, we will return to the issue of for whom managers manage that we touched on a moment ago. As before, firms as players can be seen as embodied in their boards, officers, and other policy-makers. For present purposes, we shall treat the players for society as democratically elected legislatures and executives, along with the agencies accountable to elected officials. With that as background, we can draw on the standard content of long-standing political, social, and ethical debates to tell a pessimistic Disharmony/Dilemma story about the travails of egoistic, competitive reason, which I call Exploitation.

In Exploitation, both the firm and society would be better off if they could avoid Exploiting the other in favor of Cooperating with the other. But, as always in Disharmony, one is better off taking the value-reducing path of Exploitation. Business decision-makers benefit from favoring corporate insiders and from imposing negative externalities; political decision-makers benefit from imposing electorally popular but value-destroying policies on firms. If only both of you could somehow call a halt to your egoistic, competitive Exploitation of the other—which can take place through ego-enhancing but socially destructive glorification of one's own virtue, and demeaning of the other side's ethics, as well as through pecuniary means—it would be wonderful. But how is that to happen?

As with the Slacking and Dishonesty stories, the symmetry between the players in the Exploitation story is controversial, and has its own political valence. Ordinary people and scholars on the left side of the spectrum will tend to doubt that political rip-offs of business are a real issue in a democratic system, and their peers on the right side of the spectrum will tend to doubt that market-oriented, shareholder value-guided businesses can rip off society in any serious way, as long as they are unassisted by the coercive powers of government. Centrists, especially those who believe that ethical as well as pecuniary interests are important, are therefore the most likely group to find Exploitation a congenial story. But they will, I would suggest, encounter some difficulties in trying to make constructive change based on the pessimistic, "both sides are right about the other" Exploitation story. Spreading the blame around, much as it does, in my view, represent moral progress over an us–them polarization that sees one's side as pure and the other as debased, leaves us inhabiting a grim, disturbing landscape. Further, centrist proponents of an Exploitation account of the firm–society game may be seen as falling prey—and may in fact fall prey—to their own version of Choleric us–them polarization, in which they see, or are seen as seeing, their own centrism as morally elevated, and left and right politics as morally inferior.

As a flipped alternative to Exploitation that is conducive to Sanguine reason, I would suggest a game that we can call Rationality. Here, both businesspeople and politicians are operating according to logics that generally, but not always, enhance social welfare. The firm through its representatives is understood as in fact upstanding and worthy, as is the government. But they are nonetheless in Disharmony. Why? The only way they can both achieve Highest Joint Value is if they both abandon Rationality for a Balance between System 2 logic and System 1 emotion and intuition. In so doing, they can arrive at common ground that is denied to them by their respective worthy but partial logics. But the familiar problem applies: No matter what the other does, you are better off sticking to Rationality. It would be wonderful if you could align in Harmony—but the logic of Dominance is against you both.

Through viewing the other (and oneself) in the benevolent terms of Rationality, rather than the jaundiced terms of Exploitation, progress may perhaps be made in thinking through new, Sanguine System 2 approaches to familiar, difficult problems. In particular, the classic business ethics issue of shareholder orientation versus stakeholder orientation that has divided Jensen, Friedman,[12] and others in the mainstream of financial economics from the mainstream of business ethics may be understood in different ways. Instead of skeptically viewing advocates of shareholder value as externality-off-loading Exploiters, or as apologists for them, one may engage them as Rational parties open to different possibilities, including the idea that management oriented toward enhancing the value of a diversified portfolio of all stocks, not of stock in company X alone, is both a defensible interpretation of Rationality and a potential way to solve Rational Disharmony. Similarly, instead of understanding advocates of stakeholder management[13] as nest-featherers, empire-builders, value-destroyers, or apologists for failed left economics, one may work with them sympathetically, on the premise that they are open to various possibilities, including the idea that avoiding favoritism to insiders is an important ethical component of the Rationality they embrace, and a potential way to alleviate Rational Disharmony.

A like process applies in engaging the players on the government side of the Rationality game in the development of potential Sanguine approaches to the problems of government. Instead of viewing politicians and the rest of government cynically—through one ideological lens, as Exploiters who rip off the public through ripping off business and other value-creators in favor of public employees and other insiders, or, through a different ideological lens, as Exploiters who rip off the public by being in bed with business—one may view them sympathetically, as governed by a basically Harmonious Rationality that nonetheless fails some of the time. As a practical person, one may engage "single value metric" politicians who believe that following the electorally oriented Rationality of "what it takes" is the right touchstone, and also "stakeholder" politicians who follow a Rationality that includes considerations other than electoral victory, and work pragmatically with both camps on the development of approaches that may help their Rationality Dilemmas to be solved.

A final note: My own belief is that flipping Disharmony games is a technique that is best used for Sanguine reason, rather than for competitive, Choleric reason. But the latter form of reason is important, too. To turn from political ethics back to business ethics: To the extent that partisans of stakeholder Rationality want to argue that their approach is better aligned with optimal Balance, and partisans of shareholder Rationality want to argue the same, I wish good luck and good arguments to both of them. Competitive Harmony, though hardly underrepresented in the System 2 repertory of academics in the way that I believe Sanguine Harmony is, is a game worth playing well, and in innovative ways. If a game-theoretic version of virtue ethics were to inspire creative System 2 Choleric Harmony games, that would be a very good thing, much as the objective here is the very different one of fostering Sanguine games.

* * *

Taking Sanguine Harmony Seriously

Interesting and important as Disharmony is, it is Harmony that is the master game of human nature in the conception here. In the games that follow, I want to suggest that there are two ways in which we can flip a conventional understanding of Harmony games in which the players have a shared interest in playing the HJV strategy. The first way involves failed Harmony games in which one player deviates from HJV, or appears to do so, and provokes unhappiness on the part of the other player. My main idea will be that such unhappiness can be alleviated if one flips one's understanding of the unsuccessful game from Phlegmatic Harmony, in which oneself or the other player is being dumb in failing to follow his or her own interest, to Sanguine Harmony, in which oneself or the other player, perhaps understandably, does not realize what the other wants. The second major way involves understanding successful Harmony games in Sanguine terms. Here, I want to suggest, the shift from a Phlegmatic to a Sanguine perspective may aid one in moving from stolid satisfaction to a more elevated state.

The Business and the Customer #1: From Idiots! to I Get It!

In Harmony games, the players' relevant concerns—which may be pecuniary, social, principled, or anything else—are aligned. That does not mean, though, that a Highest Joint Value outcome is necessarily reached in a Harmony game. In practice, one or more players may deviate from HJV, resulting in a less than optimal outcome. When that happens, those of us with Choleric dispositions may actually get madder (and those of us with Melancholy dispositions may get sadder) than we would if we experience the game as Disharmony. In Disharmony, a player who plays HJV is likely to understand, if not to like, the logic of Dominance followed by a player who deviates from HJV. In Harmony, on the other hand, tolerance for deviation may well be minimal. In a Choleric humor, one may feel, or say to oneself, "What an idiot!" or, if the other player is a business, "What idiots!"

Let's tell an Idiots story: You are a customer of a railroad that operates multiple lines, including a light rail line that connects two heavy rail train stations. You are used to taking the train from your home to station 1, catching the light rail a minute later to station 2, and then boarding an express train there to your final destination. The railroad has changed its schedule, so that your train leaves your home two minutes later. To your ire, when you arrive at station 1, you see the light rail starting to pull out. You chase after it unsuccessfully, pounding on the back window of the car, thinking angrily that the people who run the railroad really ought to have the schedules of their own lines aligned. Idiots!

Now, a second Idiots story that goes the other way: You are a conductor on the same railroad, which has a rule that no music is allowed unless the rider is wearing headphones. A rider is wearing headphones while listening to music. Nonetheless, another rider in the car complains, saying he can hear the music. You point to the sign to indicate that it's okay, and the complaining rider scowls. Later in the day, it happens again—another rider with headphones, and another complaint. Again, the complainer is surly when you explain. This time you're pretty ticked off, and you raise your voice on the platform to the rider as he gets off. Idiots!

The suggestion to the angry customer is that instead of thinking of the game as Phlegmatic Harmony, in which the business is failing to do what is obviously in its own interest, one can benefit from thinking of the game as Sanguine Harmony, in which the business is trying to please you and other customers, which it has real but fallible skills in doing. The suggestion to the angry conductor in the second story is similar: Instead of thinking of the customers as idiots and jerks who just don't get it, one can think of them as trying to please by upholding the railroad's policies, maladroit as they may be in that regard. In seeing the other not as blind to what is obvious, but as understandably mistaken, one can, perhaps, correct a Choleric imbalance—or a Melancholy one, if we substitute downcast protagonists for angry protagonists. You can get it yourself.

The Business and the Customer #2: From Complacency to Gratitude

It is not from the benevolence of the butcher, the brewer, or the baker that we expect our dinner, but from their regard to their own interest, Adam Smith famously observed. Equally, he could have observed, it is not from the benevolence of the shopkeepers' customers, but from their regard for their own interest, that the butcher, the baker, and the brewer expect their payment. In the Phlegmatic, calmly optimistic Harmony story that Smith told in the eighteenth century and that remains—rightly, in my view—at the heart of the ethic of business, self-interest aligned with self-interest creates social value. At the same time, there are other Harmony stories to be told. We can, in the spirit of the standard Invisible Hand story, calmly appreciate the way in which a shared pragmatic ethic of energetically following one's interests can achieve wonders of social coordination. But we can also feel a happy, even joyous, sense when we reflect upon how a particular Harmony game between a business and a customer can succeed not only because of industry and sensible calculation, but also because of a shared Sanguine, generous spirit.

Consider an out-of-town customer at a bakery that offers baked goods that are mostly sold to neighborhood residents, who are nearly all of a different ethnic group from the customer. The customer orders coffee and a pastry that reminds him of a Danish. No prices are posted; one person behind the counter wraps the pastry in paper, while another gets the coffee; the customer pays, and takes the purchases back to the table where he has left his pack. The pastry is tasty; the coffee is all right; the Internet connection is fine; the out-of-town customer works on his laptop, while local customers chat and read newspapers written in their language.

This small, successful Harmony story can certainly be reacted to and understood in Phlegmatic terms. At a System 1 level, the customer and the shopkeeper may have a Complacent, down-to-earth, perhaps slightly bored, feeling of everyone going all right, with perhaps a worried, Melancholy feeling or two—Are you the leading edge of gentrification? Will your regular customers be bothered if more outsiders with laptops start coming? From a System 2 perspective, one may think from the customer's perspective about the possibility of coming again to the bakery some time, or from the business's perspective about whether more out-of-town customers are likely in the future. But one can also flip the story, from Phlegmatic Complacency to Sanguine Gratitude. One may feel as the customer that one has been given a great gift by the business. In caring about their success, they have cared about pleasing you. In opening their door to the world, they have created for you an open window into a broader world than your own. One may feel a different, but related, joy as the shopkeeper. In the trust the customer reposes in you is a homage to you. As either player, or as both players together, you may experience a beautiful, if necessarily evanescent, moment of joyous enlightenment.

So too with System 2: In the flipped story of Gratitude, one may reflect on the technology of the Sanguine. As the out-of-town customer, one may consider how one can open oneself more effectively to the moments of joy that, fleeting though they must be, are the peak of one's life in this world. Perhaps, one may decide, paradise for you—for this month at least—may be found in stepping out of your car into neighborhood bakeries. As the shopkeeper, one may ask oneself the same question about joy and how one can experience it. Perhaps, one may decide, joy was facilitated by what you did in putting up a World Cup poster in your window . . . perhaps, you may decide, there are other good ways to allow joy, both the joy of your customers that you feel as well and the joy that you feel directly at the respect and homage of your customers, to come more often into your life.

Business and Society, Employee and Employee: Love

We now move from the bakery back to the broad domain of the relationship between business and society, both embodied in the people who are their representatives. Before, I considered that relationship in Disharmony terms, and suggested that understanding how Disharmony can arise from good qualities may help players on both sides of the business–society game to view one another more charitably. Here, we start instead on the more optimistic terrain of Harmony.

We can call the Phlegmatic Harmony game between business and society Respect. In this game, elected officials and their agents believe that the invisible hand of pecuniary self-interest drives business to feats of productivity that help sustain the society's arts, culture, government, and politics. They respect business, and businesspeople. For their part, people in business believe that different, nonpecuniary mechanisms of self-interest, such as politicians' electoral motivations, and nonprofit leaders' reputational interests, help them to do a basically, though not always, good job that provides the social grounding for businesses to succeed. They respect society, and those who work in nonprofits and government.

Respect is a very fine game. But, on at least some occasions, one might consider a flipped, Sanguine version of the game, which ramps up the cool positive regard of Respect into a much more intense feeling. Instead of regarding the other side as motivated by self-interest that works out for the best, one may see them as also motivated by love, whether for the whole, for risk, for a cause, or for certain people, that turns out for the best. In seeing them that way, one may return their love.

The flip from Respect to Love works well—better, I would suggest—if we move from the macro level of business and society back down to the micro level of employees. One may respect the other employee as, like oneself, being motivated by calculation and competitiveness that works for the benefit of all. But one may also love the other as being motivated, like oneself, by love of one kind or another.

We do not want a System 2 technology of love that leaves us in ecstasy all the time. But with reflection, we can open ourselves to moments, unpredictable and passing though they may be, of loving the other we are in Sanguine Harmony with. And as managers, part of our job—a part that makes us ethically controversial, but that is inscribed in our distinctive social role and our distinctive social duty—is to think in System 2 terms how to create System 1 feelings that redound to the benefit of the whole. In doing that part of our job, we could do worse than to reflect on how we can enhance the possibilities for Love.

The In-Between Games: Partial Disharmony and Partial Harmony

Our central game in this book is Sanguine Harmony, and our second-rank game is its doppelganger, Choleric Disharmony. But the in-between games in which there is neither a clear alignment nor a clear opposition between the ethic of Dominance and the ethic of Highest Joint Value are also important in business and elsewhere. With their mixed motivations and their asymmetries, these games are particularly interesting as bases for reflecting on the System 1 emotions involved in passivity, leadership, and followership, and the possibilities for System 2 interventions in a Sanguine spirit. In the following sections, I tell three brief stories of how standard interpretations of the games can be supplemented by nonstandard, Sanguine ones that may help people in organizations to better address the dilemmas of leadership and followership.

Partial Disharmony: Managerialism

In Partial Disharmony, recall that the best response to the other players deviating from Highest Joint Value is to play Highest Joint Value oneself. So, in the standard Chicken story of Partial Disharmony, if the other player is a Hawk, you have an interest in being a Dove (or Chicken). As with Disharmony, the first step in addressing Partial Disharmony will be to flip the standard story, to show how admirable character qualities, not simply reprehensible ones, may result in unfortunate outcomes.

For our scenario, consider the classic corporate (and also nonprofit) scenario of relations between a chief executive officer and a board that hires, oversees, and, if necessary, terminates the CEO. In standard Partial Harmony/Chicken, the danger is that the aggressive player grabs, and the other player caves. So, for example, a dominant CEO runs roughshod over a doormat board.

I would suggest the following alternative story, which I call Managerialism. In it, both the CEO and the board are motivated by a strong and, we shall assume, worthy commitment to measurable goals, accountability, and all the standard apparatuses of modern management. The board sets goals for the CEO and itself, and meets regularly to evaluate the CEO. All is well, it might seem. But in fact, the relationship is dead, and the goals are a hollow shell. The managerial spirit rules over governance, albeit in a far gentler form than in a standard Chicken story of a CEO running rampant over a docile board. Everyone is courteous, and private and public meetings are conducted with professionalism and with proper attention to legal and ethical constraints. But there is no vitality, no life, no soul in the board and in the relationship, and that hurts the organization. Everyone's motives are admirable, and yet . . .

Is there Sanguine System 2 reason that can be applied to make this situation better? How can the CEO and the board both lead, rather than merely manage? Here, the claim of this chapter about the underrepresentation of the Sanguine in modern human reason runs into the fact that a host of management consultants and authors offer a host of Sanguine ideas to try to create the gold of leadership out of leaden, soulless management, whether in the CEO-board setting or another setting. Some of the ideas are, in my judgment, very good ones, if understood not as formulas, but as keys for flipping one's situation in a way that opens the door to change. Currently, though, these ideas are seen as lying on the "pop," nonscientific side of management, and the parallel nonphilosophical, nonempirical, nonacademic side of business ethics. Management and business ethics as they are now are riven by a deep divide between what is understood as academically serious, and what is of interest to practitioners.

So how, if at all, does the moral emotions approach to game theory advanced in this book help the situation? It may be of value, I hope, at the level of helping those of us who are on the academic side of the management and business ethics divides to open our minds and hearts to those on the popular side, and helping those of us on the popular side to do likewise, so that we can move more often from mutual incomprehension and disdain—from unsuccessful Disharmony games, or from no contact with one another at all—to energetic, positive engagement—to successful Harmony games.

For those of us with strong predilections toward either the academic or the practical side—essentially, all of us who are either academics or businesspeople—moving from Disharmony to Harmony will require a strong element of Melancholy humility. To give a personal example: As an academic business ethicist, I would not presume to tell management consultant John Carver, or the devotees of his policy governance model,[14] that the game-theoretic approach to human relations advanced here provides a master key that will tell them what to do from the top down. For Carver and his devotees—including myself, for what it's worth, in my service on a local school board—I would hope for a similar forbearance—for a spirit that values abstract contemplation, even as it constantly grapples with the nitty-gritty of practice.

Equal Partial Harmony: The Flip from Principled Leadership to Principled Followership

Of all the standard game-theoretic stories, the Stag Hunt, or Equal Partial Harmony, is the one that by a considerable margin has the most inspiring moral. In this game, you and the other player can hunt for high-value stag, which requires both of you to catch, or low-value hare, which you can catch alone. The best outcome for both of you comes from collaborating to capture the stag, but if the other deviates from HJV by hunting hare, you are better off deviating, too.

In Equal Partial Harmony, a leader who commits to hunting stag, and simply continues to do so no matter what the other player does, creates an incentive for the other player to also hunt stag. Accordingly, the game lends itself to being treated as a parable of ethical leadership. By doing the right thing, and not being downcast by the slings and arrows of fortune, you create the conditions for the best outcomes to be achieved for others and for yourself.

Here, the standard story of the game lends itself well to a Sanguine spirit, which makes flipping it less important for our purposes than flipping the other standard stories is. I would suggest, though, that there is value in a flip that treats the committed player not as a leader recognized by the world as such, but as a follower, possibly a lowly one. That committed follower can lead the leader. To put the point as the title of a story: The Last Shall Be First.

Unequal Partial Harmony: The Flip from the Battle to Self-Sacrifice

We have now arrived at the last of our in-between games. The classic story of Unequal Partial Harmony is the Battle of the Sexes, in which a husband and wife who are in different locations, and cannot reach each other, have to decide whether to go to the baseball game or the ballet. The husband prefers baseball, while the wife prefers ballet, with both having an even stronger preference for being together, rather than alone. The game in its standard formulation has two Highest Joint Value solutions, and is partially Harmonious because the best response to the other player playing HJV is for you to do so also—if you as the husband know your wife is at the ballet, you go there, as well.

In the standard Battle story, and in modified Battle stories with only one HJV outcome, leadership is portrayed as an exercise in assertive egoism. Aggressive leadership gets the higher payoff. The basic idea of the flip here, as in all our games, is to assume a more attractive set of qualities than aggressive egoism, in this case sympathetic self-sacrifice, and to realize that the resultant game has challenges that are logically exactly parallel to those in the original game.

In the game that I call Self-Sacrifice, which will be discussed in relation to yeasts[15] and other organisms in the next chapter, the leader who takes the initiative gets a lower payoff, not a higher one. An example would be a situation in which a firm needs one of two employees to volunteer for an arduous, short-deadline project. The thought here is that through appreciating the reality of Self-Sacrifice, of the altruistic Leadership game and its challenges, as well as the reality of the standard Battle of the Sexes game, we can do better in crafting ways to inspire leadership and followership that draws effectively on all of the temperamental quadrants, rather than being limited to only some of them.

* * *

Alternative Stories: A Summary

I summarize where we have been so far. As noted in the beginning, my reason for the summary is that our story-telling and our story-understanding minds work through repetition. So, where we have been so far, in a nutshell: There are four main types of alternative stories we should heed, I believe. First, there are retellings of Disharmony, and Partial Disharmony, that present them as the result of positive character qualities, such as Deference, Agreeableness, and Rationality, rather than of negative qualities. If one wishes to remember one in the suite of three related alternative Disharmony stories I offered, I would offer the first one, Deference. Recall the problem of the amiable employee and the amiable manager, who fail to commit effectively not because of laziness, but because they do not want to step on each other's toes, and the suggestion that we can employ System 2 reason that encourages Assertion, and that starts from a Sanguine place about the motivations of all the players.

Second, we should understand how unsuccessful Harmony games can result from oneself or the other ineffectively but benignly trying to please, not from stupidly failing to see what is obvious. To get to that "I Get It!" moment, I began with a sharp-edged story of an angry train rider and conductor, and how they can see the other not as stupid idiots who fail to see their own interests, but as well intentioned if also maladroit. One could also illustrate the point in a gentler way, rooted in a player flipping the game she or he is in from self-blaming Melancholy ("I can't do anything right!") to self-acceptance ("It's all right!—it's hard to figure out what other people want!"). In either version, the door is opened to Sanguine ideas about managing the situation, and to being less the prisoner of one's Choleric or Melancholy consciousness.

Third, we want to realize that successful Harmony games offer us the possibility of an elevated state of joy and transcendence. A nice trip to a neighborhood bakery may be only that—but it may also be a thing of beauty, and a joy forever in recollection. We may move from Complacency to Gratitude, and from Respect to Love. By seeing the wonderful in the other and in oneself, one is transformed for a moment, and one may perhaps become transformed in a more enduring way. In that realization, and in that possibility, lies much material for the application of a Sanguine version of System 2 reason, in managerial forms as well as in other forms.

Finally, from the in-between games of Partial Harmony, we should realize that the difficulties of attaining good leadership and good followership inhere in well-intentioned and conscientious people, not simply in the temptations of bullying and passivity. The key story here, Managerialism, is of a flip from shared managerialism, in which neither the CEO nor the board leads, to shared leadership, in which both do.

In all of the flipped stories, there is a common thread: We ought to open our minds and hearts to new accounts that give us better access not only to Sanguine feeling, but also to Sanguine forms of reason that have not been central in the interpretation of standard game-theoretic stories.[16]

An Aide-Memoire: An Alternative Story Suite

Only if we keep on telling ourselves alternative stories of different kinds will we able to make them stick, and become better able to create our own Sanguine Harmony games. As another aid to learning, and as a support to possible self-improvement and organizational improvement, I wind up the discussion in this chapter with an ensemble of stories. For Harmony, Disharmony, Partial Disharmony, and Imperfect Harmony, I juxtapose standard stories of calculating, self-concerned reason with alternative, moral emotions stories that assume a mixture of feelings and motives, including shame, love, and loyalty. For some of the standard stories, and all the moral emotions stories, I use an imagined but plausible human prehistory.

First, Sanguine Harmony Games. A renowned standard Harmony story, as previously noted, is Adam Smith's Invisible Hand. Here, we can relate it in terms of two prehistoric people who both have gathered different kinds of fruit and want to trade with each other. Both are motivated, let us say, purely by self-interest based on their desire to eat different kinds of fruit, not by any interest in the welfare of the other. Their self-interest, though, helps both of them help the other and reach their best outcomes.

Now, a moral emotions version of Harmony. Imagine two prehistoric mothers who are spending the morning a mile or two apart in the woodland looking for fruit, each with their two young children. They are going to bring the fruit back to the group, where it will be shared and eaten. At the same time the two groups of mothers and children are looking for fruit, they are also talking to their children and engaging them.

Now, let's turn this basic scenario into a game, which I call Mellowness. Both mothers face a choice between a relentless and a mellow approach to fruit-picking. What makes this as a Harmony Game is that the mothers' preferences are in accord. Both, let's say, have a strong preference for a mellow approach in which they spend plenty of time talking with their children—teaching them—and taking breaks. Whatever the other mother does, both feel better about themselves by taking the mellow approach.

The reasonable prediction is that, left to themselves, both mothers will be mellow. Their interests are in Harmony, and the outcome of the game is straightforward as it relates to them. It may not be as straightforward for other members of their group who would like as much fruit as possible—a Harmony Game for the two players is not necessarily Harmony for the broader group.

One basic moral emotion that helps solve Harmony Games like the Invisible Hand and Mellowness is a Sanguine sense of enjoyment or pleasure in good things coming to oneself. Here, the moral emotions approach to game theory converges with the standard approach. Another basic emotion that does so, though, is Sanguine happiness at good things coming to others one is in relationship with. Additionally, another basic emotion that works well to solve Harmony is cool, Phlegmatic sympathy for everyone, including oneself as well as others. Other games involve more complicated, and sometimes twisted, moral emotions. But not all of human life requires such emotions. Sometimes simple pleasures and sympathies work very well indeed to get two people to good outcomes for both of them. Harmony Games such as Mellowness represent that pleasant state of affairs.

Now, a trickier type of game, Phlegmatic Imperfect Harmony. The standard story is called the Stag Hunt. It involves hunters who have to decide whether to hunt stag, a high-value prey that requires the efforts of both players to trap, or whether to hunt hare, a lesser value prey that each hunter can catch alone. The question is whether you trust the other hunter—if you don't, you're better off hunting hare instead of stag.

The moral emotions version of the Stag Hunt I'll call Follow the Rule. The two mothers and their children are again the protagonists. Instead of the choice being between stag and hare, though, the choice is between following a rule of looking only for fruit that you will bring back to the group, or breaking that rule by also looking for berries, which both the mothers and their children eat on the spot.

Both mothers, we will assume, have a preference for the two of them following the rule of sticking only to hunting fruit. That requires being a bit tough on one's children, who want to wander and eat berries, but the fruit is really good and the group really appreciates what you bring back. Also, if the other mother and their children have focused only on fruit and brought back lots, you're going to feel bad if you come back with hardly anything because you and your children have been wandering off in the berry bushes.

So is Follow the Rule—Imperfect Harmony with equal rewards for the players—just as easy as Harmony? Not at all. For one thing, you don't want to show up the other mother. If she and her children have been eating berries as well as searching for fruit, you don't want to embarrass her, and you're pretty sure the group will understand, even if they're not so happy about the smaller amount of fruit, just as they understand it in the Harmony game when the two of you spend time talking with your children rather than hunting for fruit every second.

So, you're a mile or two away from the other mother now, and it looks like there are some nice berry bushes off a little way from where you are, and your children would love nothing more than to eat some juicy berries right now . . . So what are you going to do?

Equal Imperfect Harmony games like Follow the Rule can be solved if one player can assure the other that she is committed to the action that brings about Highest Joint Value for both of them, in this case following a rule of hunting for fruit only.

One key emotion that can help solve Equal Partial Harmony games is Phlegmatic calm. If one simply persists in doing the right thing, the other has every reason to do it, too. Melancholy shame and Choleric self-punishing anger can also work. If a player in Follow the Rule internalizes strongly within herself the rule that she will only hunt for fruit because it is shameful not to, and follows that rule, the other player, whether she feels that internalized shame, or guilt, about the rule, has a very good reason to hunt for fruit as well. She does not want to feel the painful social shame that will result if she returns with a small amount of fruit while the other, rule-following mother returns with bushels.

With the third type of game, Choleric Disharmony, the ethical issues continue to be tricky. The standard story is the Prisoner's Dilemma. In it, you and a fellow prisoner accused of committing a crime have been isolated in different cells by the prosecutor. He wants you to confess and rat out the other prisoner. If both of you cooperate by hanging tight and remaining silent, the District Attorney only has enough evidence to put both of you away for six months. But there is a disturbing logic that leads you to confess: If you rat the other prisoner out, you will do better regardless of what he does. If he rats you out, too, at least you won't take the hit while he walks. Instead, you both will serve a five-year sentence. If he unlike you is a stand-up guy, you then get to walk while he rots in prison for the rest of his life. So it looks as though you should confess. He will reason the same way, and you both will serve a five-year sentence. But if only you could cooperate and get just six months . . .

In place of the standard Dilemma story and its appealing to some, alienating to others focus on cool, amoral calculation by unpleasant protagonists, we can illustrate Disharmony games with a different story of prehistoric life. We can call it Stranger Mother.

In the Stranger Mother scenario, as in Mellowness and Follow the Rule, two mothers and two children are once again hunting in the woods for fruit to bring back to the group. But now the situation is different. Instead of the other mother being someone in your band you know very well, she is a member of another band your band has met up with, and who is hard for you to understand because she speaks a different dialect from you. You are hunting for fruit with one of your young children and one of hers, and she is doing the same. The two children can learn from and understand one another much faster than you and the other mother can. In fact, that's a reason for the children being split up. As in the earlier social games, you are bringing fruit back for the group. But now the meal is for a combined group, including the strangers who are members of the other band, not just your band.

As far as you are concerned, the best thing for you and the other mother is if both of you are business-like about the hunt, with you and the children collecting plenty of fruit for everyone to enjoy. You are basically sure from your meeting her briefly this morning that she feels the same way. It is tricky, though, because the social expectation is that no one should bring back too much fruit—part of the script for the bands getting together is that the elders will chuckle at how little the mixed groups of children and their mothers have brought back. Given the norms, you will look bad to the people in your band and the other band, and you will feel bad, if you come back with lots of fruit, and the other mother comes back with only a little. If she comes back with lots, you will feel good if you do, too—but you will feel even better if you follow the script by only having a little.

What should you do? You and the other mother both do better if you are busy and business-like . . . but maybe you should be mellow and not push her child and yours . . . after all, no matter what she decides to do, you will feel better that way. But something feels wrong about that. After all, your two bands are getting together to share food. Both you and Stranger Mother want to do a good job at getting the fruit, and you both will be better off if only you do that. What to do?

This book takes an optimistic perspective on people's ability in the past and present to solve social games effectively in their day-to-day interactions. At the same time, it should be acknowledged that Disharmony games like the Prisoner's Dilemma and Stranger Mother that oppose Dominance and Highest Joint Value are hard—the hardest games, in fact.

The constellation of moral emotions that helps solve Disharmony games has multiple stars. Choler at a player who gets it wrong for the group has an important role. Another prominent star in solving Disharmony can be a Phlegmatic, calm, serene state, as limned in classical temperament theory, especially in its South Asian and East Asian variations. In such a state, the mothers in Stranger Mother may become detached from their self-involved and group-involved attachments to the results of their action, in a way that allows Disharmony to be solved as well as may be.

The next category of games, Partial Disharmony, is represented in its standard version by a story of two teenage boys playing a game of Chicken, in which they drive head-on toward one another. The best ego-boosting outcome for an individual player comes from being a Hawk and not swerving; by comparison, being a Dove, or a Chicken, by swerving, is not good. But being a Dove or a Chicken is better than the outcome if neither player swerves—two dead Hawks.

For a moral emotions version of Partial Disharmony, let's go back to two mothers within the band who are picking fruit. Now they are picking it together, along with their own two children. In this version of our basic scenario, we have a tough game, which I call Deference, using the same title, and the same logic, as for our key initial story of two workers in a modern office. The best results for both mothers come if they both assert and are leaders, both in taking care of the children and in gathering the fruit. Each mother wants to defer to the other. But if they both do that, they and their group both have a poor outcome.

Partial Disharmony games are not easy. Melancholy yielding results in a fairly good, but not best outcome. In flipped, pro-social Partial Disharmony games, one moral emotion that can help is a visceral, System 1 Choler felt by toward a player who will not Assert. To avoid that righteous anger, the passive, deferential player may instead lead. To state that possible solution is to also state the trickiness of the situation: Righteous Choler may not be righteous at all; the deferential player may be in the right practically as well as ethically, and the angry player may be a passive bully, not a genuine leader.

The final category of games to be illustrated here are unequal Imperfect Harmony games, or leadership games, in which the Highest Joint Value outcome is better for one of the players than it is for the other player.

The standard story of an unequal Imperfect Harmony game, as we have discussed, goes by the title Battle of the Sexes. In it, a husband and wife who are in different locations and cannot communicate with the other have to decide whether to go to the fights or the ballet. Both prefer being together at either activity to being alone at either one, but the husband prefers that they be together at the fights, while the wife prefers that they be together at the ballet.

Our flipped, moral emotions story to illustrate Leadership games could also be called Battle of the Sexes. To help differentiate our story from the standard one, we call it Sensitive Boy.

In Sensitive Boy, the two players are the mother we have seen in the earlier stories and the father of her young son. All three have a very close relationship with one another. The boy hunts for fruit with his mother, and also has started recently to hunt for small game with his father. He is very good at fruit hunting, and loves to talk with his mother and with girls. Although he is young, he does not seem to be as good at hunting as the other boys his age, and he is awkward around them.

Both the mother and the father want to be in agreement on how to raise their child. The mother's best outcome is if the father agrees with her on raising their son in a way that involves him being less involved in hunting than the other boys are, and more involved in activities with her, other women, and the few men in the band who associate mostly with the women. The father's best outcome is if the mother agrees with her on his trying to have their son become more like most of the other boys.

In this "no obvious right answer" situation, which is typical of Unequal Imperfect Harmony games, a number of moral emotions may come to the fore and help the players reach a solution. One is Phlegmatic calm. Hard as it may be to discern which approach is better, both parents can converge in a sense of peace about the outcome. Another, very different emotion that can help solve the game is Choleric righteousness in its System 2 form. Righteous feeling may be manifested as self-righteous System 1 anger that turns the mother and father against one another—"You just don't understand him the way I do!"; "Oh yes I do—I'm his father, and he's a boy, not a girl!" But righteousness, including its angry component, also has the potential to go along with the mother and the father talking together, expressing their emotions together, reasoning together, and most of the time drawing on their useful, irritating Choler to rear their son in Harmony.

The Heart of the Matter

The stories that have been told in this chapter, including those in the story suite about prehistoric life, are the heart of this book. In one form or another, they have lived in my mind for years, and I hope that they can live in yours. I very much include the classical stories of prisoners, teenage daredevils, hunters, and so on, as well as my alternative stories of the customer and the shopkeeper at the bakery, the mothers at a group feast, the angry train rider, and so on, as stories that I hope can live in us. Only if we can keep both kinds of stories alive inside us, I believe, can we get closer to the whole of who we are, and of what we might be. Either kind of story alone misses the point.

In this chapter, I have stressed that alternative, Four Temperaments stories can open us to Sanguine feeling and reason in a way that conventional game-theoretic stories often fail to do. Much as I believe in that interpretation, I want to stress that the opening to the Sanguine is a personal, subjective, indeterminate one, not one inscribed in ineluctable law or logic. For many years, including my years as a junior faculty member writing articles on alternative interpretations of economic modeling, I experienced flipped stories of the Prisoner's Dilemma and other models not as doors to Sanguine feeling and reason, but instead as Melancholy, Choleric tools to undermine complacent, Phlegmatic faith that our particular ethical and political beliefs are the right ones. I still find that debunking, ashamed, Choleric interpretation of moral emotions story-telling to have value, much as I now find myself more drawn to the Sanguine interpretation presented here. In the last two chapters, I return to the debunking interpretation, and to how both it and the Sanguine interpretation may help us do better in our practice of business ethics.

Central as the stories told in this chapter are to this book, they are not our stopping point. In the suite of alternative stories of prehistory I just told, I hinted at how moral emotions can help solve the difficult games—Disharmony, Partial Disharmony, and Partial Harmony—effectively. In the next chapter of the book, I discuss how that could work. I also suggest that the moral emotions that allow people to solve social games have their parallels in nature. If the exploratory argument, or conjecture, is right, the logic of the four temperaments is not only the logic of social interactions among people. It is also the logic of the evolution through interaction of everything there is and of everything that could be: the logic of dogs, fish, trees, and bacteria, not only of people, and of atoms, rocks, and corporations, not only of living things (Figure 3.1).

Type of game	Active/Yang—Dominant Strategy !	Reactive/Yin—No Dominant Strategy . . .
Positive— One good turn deserves another :)	**Sanguine/Harmony Games** I Get It!, Gratitude, Love, Mellowness	**Phlegmatic/Imperfect Harmony Games** The Last Shall be First, Self-Sacrifice, Follow the Rule, Sensitive Boy
Negative— One good turn does **not** deserve another :(	**Choleric/Disharmony Games** Deference, Agreeableness, Rationality, Stranger Mother	**Melancholy/Partial Disharmony Games** Managerialism, Deference

The standard stories of game theory illustrate the dilemmas of rationality and morality encountered and created by players motivated by the very real egoistic parts of our complex selves. The alternative, or flipped, stories of moral emotions game theory shown above and described in the text illustrate the dilemmas, and also the pleasant and joyous situations, encountered and created by players motivated by the also very real socially-oriented parts of our multiply divided selves that feel emotions such as shame, guilt, sympathy, respect, and loyalty.

Figure 3.1 Alternative Stories

Summary

We can access a Sanguine form of reason through a flipped, or alternative, form of game theory. In this approach, we assume that human beings have pro-social motivations that include altruism, empathy, sympathy, benevolence, shame, guilt, deference, loyalty, rule-following, and principle-following. We then proceed to analyze logically the very real difficulties that pro-social players of games experience in games with other pro-social players.

(continued)

> *Summary* (continued)
>
> In contrast to the egoists of traditional game theory, the pro-social players in flipped stories are not blamed by human moral intuition for their practical failings. That makes it possible for practitioners of the flipped form of game theory to employ System 2 reason in a Sanguine, happy spirit, as well as an equable, Phlegmatic one.
>
> One family of flipped game-theoretic stories, including Deference, Managerialism, and Self-Sacrifice, help us understand, and respond with calm and effectiveness, to the conundrums of pro-sociality in difficult games. A second family, including I Get It!, Gratitude, and Love, help us understand, appreciate, and deepen Harmony in our lives.

> *Exercises*
>
> 1. Pick a favorite literary novel or short story. Go to a scene in the work that you like. Consider whether the scene can be interpreted in terms of difficulties that arises for characters, or "players," with altruistic and other pro-social feelings. If not, try the same thing for another scene.
> 2. Pick a favorite movie or work of popular fiction, and do the same thing. Is there a moral lesson in the scene about the pitfalls of pro-social emotions like deference, loyalty, and sympathy?
> 3. (a) Do the exercises on judgments of intention, or blame, that are in the appendices, and are discussed further in Chapter Five; (b) Read and discuss, or reflect on, the article by Anatol Rapoport referenced in the endnotes on major types of games; (c) Read and discuss, or reflect on, passages from Schelling and Mathiesen on the altruist's dilemma; (d) Read excerpts from Eric Berne's *Games People Play*; discuss, or reflect on, the similarities and differences between his games, based on a parent–adult–child version of the superego–ego–id framework, and the games of the text.
> 4. (a) Pick a business case with an ethical dimension, preferably one that is usually understood in terms of people's antisocial or egoistic motivations. Reinterpret the case in terms of difficulties with pro-social motivations such as altruism. (b) Watch videos of, or read passages from, popular works of psychology, management, and ethics; good sources are the Chicken Soup series, Steven Covey's *Seven Habits of Highly Effective People*, and Dov Seidman's *How*. Discuss, or reflect on, the similarities and differences between these approaches and the approach in the text.

CHAPTER 4

Bringing Telos Back

A difficulty presents itself: why should not nature work, not for the sake of something, nor because it is better so, but just as the sky rains, not in order to make the corn grow, but of necessity? . . . Wherever then all the parts came about just what they would have been if they had come be for an end, such things survived, being organized spontaneously in a fitting way; whereas those which grew otherwise perished and continue to perish, as Empedocles says his "man-faced ox-progeny" did.

—Aristotle, *Physics*, Book II

It is absurd to suppose that purpose is not present because we do not observe the agent deliberating. Art does not deliberate. If the ship-building art were in the wood, it would produce the same results by nature.

—Aristotle, *Physics*, Book II

We moderns think we know important truths about the universe and its workings that our classical predecessors, no matter how brilliant, did not. In many respects, our confidence is warranted. We know the velocity with which our planet is revolving around the star that spawned it, and the speed with which it is rotating on its axis; we know the velocity with which the arm of the galaxy in which our solar system is located is turning around the center of the galaxy; we know the red shift that allows us to calculate how fast our galaxy and the rest of the universe are now pushing apart, some thirteen billion or so years after the great expansion, or "Big Bang," that got our universe going. We have a picture of our corner of the universe that is not only beautiful—the green and blue ball of Earth; the middle-aged yellow Sun that will one day grow old, swell up into a red giant, and then die; the billions and billions of stars in the great twin spiral nebulae of the Milky Way nebulae and its neighbor, Andromeda—but that accords with verifiable empirical truths in a way our ancestors' pictures did not. Further, we now know enough about how evolution works in a variety of organisms both to appreciate Aristotle's perspicacity in the first quote, in which he anticipates the essence of Darwin's theory of natural selection, and to suggest to him that he may well have radically overstated the case against favorable variations arising by chance.

Now, to the other side of the ledger: There are three connected ways in which I believe Aristotle and his contemporaries were considerably closer to a fusion of truth, beauty, and goodness in their understanding of nature than we are. First: To a much greater degree than us, they saw the stuff of ethics in wood and iron, in leaves and beetles, not only in humans, or in animals like monkeys that resemble us. They took ethical monism—the idea that ethics pervades everything in the universe, rather than inhering only in some things, notably human beings—seriously as a rational belief system, rather than as an outcome of a leap of faith. Second, and relatedly: To a much greater degree than us, they decentered human beings as the ethical omphalos around which everything else in the universe revolves. Before Copernicus, they were ethical Copernicans. Third, and the most important for our purposes: Their belief in telos went along with a Sanguine System 2 spirit of happy reason that was a highly valuable alternative to Melancholy, Phlegmatic, and Choleric reason.

How about us? Most of us are not Richard Dawkins.[1] We are wary of his sharp-edged claims that science, rightly understood, indicates that the universe is without moral purpose; we sense the imbalance toward righteous Choler in his acerbic acuity. But we also lack the ethical monism and the ethical Copernicanism of our ancestors in Greece and elsewhere. The suggestion here: Instead of getting irate at his ire, let us be grateful to Dawkins, and to others of his disposition. They are canaries in our moral coal mine. Compared to our ancestors, all of us, not just a few gadflies, are Cartesian dualists and ethical Ptolemaians. We see ethics as inhering in us, not in what is outside of us. We lack a rationally grounded sense of connection between ourselves as ethical beings and the rest of nature. The System 2 reason embodied in the science that is one of our culture's signal feats has deprived us to some substantial degree of our ancestors' belief that every creature, every human-created thing, and everything in nature has a purpose.

Most of us are also not Alasdair MacIntyre. The moral universe we inhabit does not seem as profoundly disordered as he says it is in his Melancholy masterpiece, *After Virtue*; we feel pride as well as concern about modern morality, and we doubt that the Greeks or anyone else really had a better ethical understanding all in all than our own. We sense an imbalance toward submission and shame in MacIntyre's portrait of moral collapse, and possibly also in his idealization of Aristotle and Aquinas, and in his former idealization of Marx. At the same time, he, like his opposite Dawkins, is us—and for those of us who are philosophers at least some of the time, that is doubly true. *After Virtue* has been a significant influence on this book. But instead of following MacIntyre's path of Melancholy brilliance, the thought here is to try to make progress in a different way: We can, I have suggested, help ourselves to become ethically unstuck by applying a version of game theory that gives us better access to Sanguine reason, and in so doing provides an alternative to the Melancholy, Choleric, and Phlegmatic forms of reason that are the reigning spirits of our age.

The aim in this chapter is to extend the focus on human beings in the earlier parts to other animals, other organisms, inorganic matter, and to created entities, such as firms. I suggest that, although the particular emotions and behaviors associated with the four temperaments are not universal—a yeast does not feel anger, or self-consciously calculate its best response to the behavior of another yeast—the temperaments, understood as payoff functions that enable parts to solve games on their behalf and that of wholes, are indeed universal. In the exploratory case to be made here, the correspondence David Sloan Wilson notes between the languages of evolutionary biology and ethics, with their common reliance on concepts such as altruism and punishment,[2] is not a coincidence, or a necessary artifact of human consciousness that is divorced from material reality. It is a reflection of a union of ethics and reason in all creatures and all things in Nature, including us.

Back to the master: In a dialogue with us moderns on evolution, I believe Aristotle could stand his ground on the issue of how natural selection works once variation arises. Specifically, he could argue that there is a purposive ethical logic, a telos if you like, in natural selection, in the success of the ox-faced ox-progeny, and the failure of Empedocles's man-faced oxen. Living creatures, but also things, such as molecules and corporations, that solve social games successfully by their own lights, and also by those of a whole of which they are part, flourish relative to those that are deficient, or excessive, in their attachment to the part, or to the whole. So, I believe, he would say; so, at any rate, I suggest in this exploratory part of the book.

An advisory: This is the most technical, and at the same time the most speculative, part of the book. Some of the arguments that follow in this chapter may try some readers' patience. If one finds oneself reading them for the poetry rather than the logic, that is fine. The logic, right though it is to the best of my judgment, is advanced in a suggestive fashion, with the hope that it can be developed, interrogated, and improved in the future by scholars who have logical and mathematical reasoning skills that are greater than, or in any event different from, my skills in those domains.[3]

* * *

The Evolution of Game Theory

Game theory is no longer tethered to its post–World War II origins as a method of analyzing choices made by human actors who are expected to be rational calculators of odds and advantages, such as Sherlock Holmes and his adversary Professor Moriarty, cost–benefit analysts, poker players, Wall Street traders, and military planners. That original, or "classical," approach to game theory remains highly influential and important in economics, business, and elsewhere. But starting in the early 1970s, a very different, evolutionary approach to game theory began to appear, with the 1973 *Nature* article by Maynard Smith and Price that introduced the concept of an evolutionarily stable strategy serving as a milestone.[4] The evolutionary approach now rivals, and possibly exceeds, the classical approach in its overall influence and significance. Under the evolutionary approach, game theory is a method of analyzing all actions or behavior, including the behavior of slugs, rabbits, genes, galaxies, and of human beings acting emotionally or intuitively, rather than through reasoned calculation.

In evolutionary game theory, a "generous genes," altruistic, collective maximization perspective on evolution is logically intertwined with, and coequal to a "selfish gene," egoistic, individual maximization perspective. The basic reason for the parallelism and parity of altruism and egoism is very simple. Organic and inorganic matter from very small to very large—from atoms to genes to yeasts to people to stars to the universe, to a possible multiverse of interacting universes—consists both of parts and of wholes. We can view the actors in any given game—two one-celled organisms, say—as separate, independent individuals. In doing so, we logically adopt the proposition that "selfish bacteria" strategies that maximize the adopters' reproductive fitness will be favored by evolutionary logic. But we can also view each bacterium as a whole, with its parts subject to a logic of the whole. In doing so, we also logically adopt the proposition that "generous nucleus, generous mitochondria" strategies for the dependent parts of the bacteria are what maximize the fitness of the parts, and will be favored by evolution. That is not the end of the pairing of egoistic and altruistic games, to be sure. If we in turn view the nucleus and the mitochondria as separate actors, we recreate the selfishness game. But nested within that second game of egoism are two games of altruism, of the parts of the nucleus and the mitochondria and their incentives to act altruistically. Those altruistic games in turn yield to egoistic games of the parts each viewed as individuals, and to a further altruistic story, and so on.

Altruistic evolutionary games, just like the egoistic evolutionary games with which they are twinned, are characterized by the logical possibility of Disharmony. In the individual-centered, selfish bacterium account, two separate bacteria will do better in some situations if only they could figure out a way to cooperate together—they are locked in egoistic Disharmony. Equally, in the matching account of generous parts, the nucleus and the mitochondria will do better in some cases if only they could figure out a way to Assert, to avoid the crushing social pressure of the cell in which they are confined and both go their own separate ways—they are locked in an altruistic Disharmony of Deference.

As large, highly complex organisms made up of a huge number of component subparts, it is understandable, and appropriate, that human beings deeply value the altruistic aggregation of parts into a complex whole. Without the altruistic "parts acting for the whole" side of evolution, we could not exist. But the reverse is true as well. Without the egoistic "individuals acting for themselves" side of evolution, we would be no more than the captive mitochondria and nucleus. Just as we should be open to the normative possibility that our human world is oversocialized and overaltruistic, as well as to the possibility that it is undersocialized and overegoistic, we should be open to the parallel normative possibilities in reflecting on the universe, and on its possibly overorganized, oversocialized as well as its possibly underorganized, undersocialized nature.

In addition to viewing the inorganic as well as the organic universe in terms of altruism and self-concern, it is also possible, and for that matter called for under the logic of evolutionary game theory, to view it in terms of other sentiments as well. The nested view of inorganic and organic matter, as consisting of both parts and wholes, calls for a complex ethical repertory. To understand nature and ourselves in terms of individuals entails the idea that individuals may succeed in evolutionary terms not only through egoistic, *homo economicus* strategies, but also through competitive strategies that focus on beating one's rivals. The nineteenth-century and early twentieth-century Social Darwinist interpretations of evolution[5] in those terms, politically and ethically tendentious as some of them were, captured one real part of the moral logic of evolution. Individuals, from the microscopic level of elementary particles all the way up to the macroscopic level of universes in a multiverse, do indeed have an incentive to experiment with strategies that focus on harming their evolutionary rivals, as well as helping themselves. In doing so, competitive individuals face particularly acute Dilemmas. If only they could both forbear from harming the other, they would both be better off. But no matter what the other player does, one is better off if one takes a bite out of the other. So goes Choleric, competitive Disharmony, which is akin to, but far more potent than, the egoistic, Phlegmatic Disharmony of individuals who are indifferent to others.

On the other side of the twinned ethical logic of evolution from Choleric Disharmony is Melancholy Disharmony. It applies to parts of wholes, and is the companion of the altruistic, other-oriented logic of the part. The basic idea is as follows: In programming parts like the mitochondria and the nucleus with an evolutionary logic that makes them faithful servants of the greater good, it is desirable if possible to imbue them with "shame," with this unconscious version of shame being defined as a negative payoff for actions by the part that deviate from the welfare of the whole. In other words, to make the logic of the whole work as well as may be, it is helpful if the mitochondria and the nucleus that stray from it feel a twinge, or a sharper pain, that helps keep them in line.

Like all the other parts of the logic of evolution, shame, useful as it is, has its Disharmony as well as its Harmony. Melancholy Disharmony at the cellular level can be expressed as follows: The mitochondria and the nucleus could do better if only they could both dissent from the social logic of the cell. But no matter what the other does, one is better off acquiescing. It is too bad when altruistic mitochondria and nuclei cannot solve Sanguine Disharmony, and wind up losing value they could have had if they had Asserted and acted for themselves. It is much more painful when they cannot solve Melancholy Disharmony, and wind up suffering—a suffering that we cannot understand, to be sure—from their inability to dissent from a norm that works for the whole but not for them. To put the point more generally: Shame is to altruism as competiveness is to egoism. Just as competitiveness is a ramped-up, valuable but also highly dangerous—one may resort to violence rather than competition within the rules of the game—way of enforcing the egoistic logic of individuals, shame is a ramped-up, valuable but also highly dangerous—the part may lose individuality altogether—way of enforcing the altruistic logic of wholes.

In human beings and in creatures similar to us, shame is a self-administered punishment that feels different from external, socially administered punishment. At the level of mitochondria and nuclei and certainly at that of inorganic matter, that distinction, we may safely assume, loses its experiential meaning. That means that the point of the last paragraph about the parallels between egoism and competitiveness and altruism and shame can also be stated in terms that substitute punishment for shame: In evolutionary terms, punishment, like shame, is a highly valuable, highly dangerous way of enforcing the logic of the whole.

The Four Temperaments as Universal Devices to Solve Social Games

Under the worldview of evolutionary game theory, within us, just as within all other living and inorganic matter, there is an It as well as an I. But the It within our emotional, intuitive human moral selves is not Freud's irrational, passionate id, but the impersonal logic of evolution. Instead of thinking of ourselves as a rational Dr. Jekyll on the outside and a savage Mr. Hyde within, we do better to think of ourselves as Harmonizing social butterflies on the outside, with the beautiful, implacable, universal, all-knowing mind of evolution, aka Nature, inside. Although we as human beings have no clarity as to whether the game we are playing is Harmony or Disharmony, the probing logic of evolution that is universal within us and everything else us will figure it out. A perfect information game is counterfactual for human beings, but not for Nature.

We create Harmony games with our social butterfly human peers. Not all species, and certainly not inorganic matter, have our impressive capacities in that regard. But evolved Nature, and evolved Nature within us, is, we conjecture in this section, quite able to solve the difficult social games. The basic idea to be explored here is that the four temperaments to which this book is devoted are not simply human phenomena, but are universal, evolved devices to foster interactions between parts that are beneficial both for them, and for the whole in which they are contained.

In the picture of the universe to be painted here, Dawkins has part of the truth: There is indeed selfishness in our genes, as well as in us. But reverse Dawkins also has part of the truth: There is altruism everywhere in Nature, from self-abnegating mitochondria all the way up to human beings. That is, both the self-absorbed and the social part of the Sanguine are everywhere, not just in us. More: There is Melancholy shame everywhere, not just in human beings, in the form of useful mechanisms for self-punishment on behalf of the whole. Further, there is Choler everywhere, not just in human beings, in the form of useful punishment on behalf of the whole, and also in the very different form of useful competitiveness on behalf of the part. Finally, there is Phlegmatic detachment everywhere, not just in human beings, for the ability to calculate a best response and to mutually align in playing Highest Joint Value, rather than be governed by unchangeable programming, is very helpful for nonhuman as well as human players to have.

As a hypothesis, or conjecture, about how the humors can combine in human and nonhuman players of games to achieve Highest Joint Value, I propose, or conjecture, the following: Individual players of all kinds, of all games, are divided into two subselves for purposes of action, and two more subselves for purposes of calculation and response. The first active subself is a positive Sanguine quadrant that values both its payoffs and those of the other, and the whole. The second active subself is a Choleric, competitive, righteous quadrant that negatively values others' payoffs when they exceed its own, and that inflicts punishment by negatively valuing its own payoffs and others' payoffs when they violate norms of the whole. Now, the response side: The first reactive, responding subself of the player, human or otherwise, is a Phlegmatic quadrant that blends calculation and detachment. The second responding subself is a Melancholy quadrant that blends rule-following and self-punishing Shame at violating norms.

The basic idea here is that human beings and other objects in Nature with a Sanguine acting subself, a Choleric acting subself, a Phlegmatic reacting subself, and a Melancholy reacting subself do well in attaining Highest Joint Value when playing difficult games with similar objects. The core intuition behind the argument can be stated as follows: One of the two acting subselves in the player, either the Sanguine or the Choleric, will have a Dominant strategy of playing HJV in all of the difficult games. Given that, one or both of the reacting subselves will tilt toward playing HJV. The calculating, Phlegmatic subself will tilt that way in cases in which it might not "want" to otherwise because of the prospect of being punished, while the Melancholy subself will do because of rule-following, self-punishing Shame.

In what follows, we will develop that intuition about a subself with a Dominant strategy of playing HJV for different major kinds of difficult games. In the different games, different temperaments and subselves step to the fore, as we will see. In all the difficult games, the dark-side, Choleric subself, not just the bright-side Sanguine subself, is important to attaining HJV. Similarly, in all of them, the reacting Phlegmatic and Melancholy subselves, not simply the engaged, acting subselves, are important in reaching HJV.

In the informal evolutionary models that are presented here, we are interested in the individual actors as parts of the whole. We are trying to find out the nature of the programming in them that could lead to an outcome in which HJV is achieved in a sustainable manner, that is, one that is not vulnerable to incursion by pure egoists. In other words, we are back-solving a given individual to find out a plausible automatic, programmed, instinctive, or intuitive set of behaviors that result in HJV.

Phlegmatic/Imperfect Harmony Games with Unequal Outcomes

For our first game, let's consider one that is distinctive, and distinctively relevant to business and to business ethics, in its involving unequal outcomes for the players when they both play their Highest Joint Value strategies. Both players do well relative to the situation in which they do not coordinate, but one does better.

As always, we need stories. Let's consider three versions of Unequal Imperfect Harmony. In our first story, Social Leadership/Followership, achieving Highest Joint Value involves collaboration between for-profit firms and nonprofit and governmental organizations in providing services for the homeless. The terms of the collaboration either put the for-profits or the not-for-profits in the Lead, with the other sector as the Follower. There is a payoff inequality between the Leader and Follower positions, which could favor either the Follower, or the Leader. In the second version of Imperfect Harmony, Self-Sacrifice, discussed in another form in the last chapter, achieving Highest Joint Value involves getting some yeast cells in a simple, aggregated multicellular "snowflake" yeast[6] to "Lead" by self-sacrificially undergoing apoptosis—dying—while other cells "Follow" and survive. The cells that Lead and die do better than if no cells died, since their genetic material gets passed on at a higher rate—but they do not do as well as the cells that Follow. For our final story, recall Sensitive Boy, in which the mother and the father want to coordinate on raising their son, but have different ideas of what is best for him, them, and the group. Here, Leadership is on the face of it advantageous relative to Followership, though the advantage may be reversed in practice when one takes into account the fact that the parent who gets his or her way may have to do greatly more work than the parent who does not.

A bright side of solving an Imperfect Harmony game, such as Self-Sacrifice, Sensitive Boy, and Social Leadership/Followership, is that given the other player is playing HJV, HJV is your own best response. Phlegmatic egoism is not a problem, as it is in Disharmony Games. But the Phlegmatic, reactive subself needs something to respond to in the other player. It needs to respond to a part of the other that has a Dominant strategy of playing HJV by either leading or following.

In Unequal Imperfect Harmony, the "hero temperament" that provides a stable, Dominant anchor that allows the game to be solved is a somewhat surprising one: competitive Choler. Let's see how that works.

Suppose you are the player—the parent, the cell, or the firm—who should sacrifice for the greater flourishing of the other and the whole. You can rely on the Choleric subself of the other player playing HJV. The competitive subpart of that Choleric subself wants to take the self-aggrandizing path. In that competitive "desire," it is not countered by the self-punishing, Guilty side of the Choleric, because the other player's self-aggrandizing is in accord with the whole, and hence is not countered by a social norm.

Given that the Choleric part of the other player is going to act reliably in a self-aggrandizing way by playing HJV, your calm, calculating Phlegmatic subself and your Melancholy, norm-following, Ashamed reacting subselves converge on your strategy of self-sacrifice (the HJV strategy for you). You either Lead or Follow, whichever is the more sacrificial (though still beneficial to you) thing to do, if you are the person or the firm. So, for example, assume a game in which HJV comes from a nonprofit homeless shelter acting in a self-sacrificing way by Leading in providing services for the homeless, while a for-profit firm acts in a self-aggrandizing way by Following and not providing services. In this game, Nature in the form of the self-aggrandizing, competitive, Choleric side of the firm will reliably defer, which will signal the shelter to Lead in providing the services. On the other hand, if we assume HJV comes from the shelter aggrandizing *itself* by providing the services, while the firm acts self-sacrificially by deferring, then the firm can rely on the self-aggrandizing side of the shelter to lead by providing the services. Either way, HJV is reached by perfectly informed subselves.

I hasten to acknowledge that the perfect information game just described, in which the firm and the shelter lead and follow in order to achieve Highest Joint Value, is very far indeed from the real world of unclear payoffs. As a real world player, one lives, or exists, in a fog. The firm is an abstract entity that knows nothing. You as a conscious human actor acting for the firm may know what is self-sacrificial or self-aggrandizing for it in immediate financial terms. But as business ethicists are correctly at pains to point out, that knowledge of one form of the firm's immediate payoffs is not equivalent to knowledge of its total, overall, forward-looking payoffs, or of the other player's payoffs.

The point of the perfect information exercise is not to claim that daily reality is guided by a Panglossian logic that leads the HJV immanent in the fusion of the parts and the whole to be attained in particular games. In reality, the humans involved in the firm and the shelter will not know the game they are playing, and may well not successfully solve an Imperfect Harmony game between their organizations, if indeed that is the game they are playing. Rather, the point of the exercise is that Nature favors the survival of parts and wholes—for-profit firms, shelters, and societies in which they are embedded—in which the parts and the whole have a particular makeup that helps them reach HJV.

The same point about real world obscurity, conjoined with an abstract perfect information logic of Nature that favors a makeup of the players that leads to stable HJV, applies to the Sensitive Boy and yeast stories. The mother and father in Sensitive Boy do not know in practice whether rearing their child in the way favored by her or by him will lead to the elusive Holy Grail of stable HJV. Nor do they truly know what is self-sacrificing and what is self-aggrandizing, when payoffs of all kinds, pro-social as well as egoistic, are taken into account. They may well in practice fail in solving their Imperfect Harmony game, if that is indeed the game they are playing. But here, as with the firms, Nature favors mothers and fathers with the emotional makeup that gets them to HJV more successfully than their peers with other makeups. So too with the nonconscious yeasts. They know nothing, and reason not. But Nature favors a makeup in them in which the yeast that should sacrifice by going apoptotic does so, with the other yeast benefiting from its sacrifice.

Choleric/Disharmony Games

For our business ethics case, recall the Rationality story of organizations that have a Dominant strategy of following System 2 calculation and deontology, even though they would do better if they could both move toward a better balance between System 2 reasoning and System 1 intuition. For our biological story, let's use an account of parts of a cell that have a Dominant strategy of remaining together, even though separation would be better for them both—we can call it Locked-in. Finally, recall the Stranger Mother story of the mothers who are impelled by social pressure not to collect too many fruits and berries, even though they would both do better if they worked harder.

Now, the Disharmony analysis, focusing on the firms in the Rationality story. The Sanguine subself is the first hero here. With its aggregation of self-concern and sympathy, it has a Dominant strategy of playing the HJV strategy by blending in more System 1 intuition. Sanguine would "like" the other firm to play HJV, but it will do so itself, regardless of what the other cell does. How about Choleric? It's not as upstanding, but it's not all bad. Its competitive side wants to beat the other firm by playing the non-Highest Joint Value strategy. Choleric doesn't have a Dominant strategy of doing so, though, because its self-punishing side puts a negative value on playing the non-HJV strategy if the other firm plays HJV.

Now, stage 2 of the analysis, in which both firms "consider" their best response to the other. Here, it is the Melancholy reacting subpart of the firm that becomes the hero. It "calculates": My Ashamed side wants to play HJV here, given the possibility the other firm has done so—so I do.

Finally, stage 3. Here, punitive Choler directed on behalf of the whole is the hero. The Phlegmatic reacting subpart of the firm—which here, as in all the other cases, one may see as embodied in the individuals who act for the firm—wants to play the suboptimal Dominant strategy. So does a hypothetical pure egoist, who can invade a population that has solved Disharmony successfully. What keeps the Phlegmatic subpart of the firm, and the hypothetical purely egoistic invader, in check? Punishment, or the prospect of it.

Here, the four temperaments analysis of Disharmony converges with the conventional wisdom about the Prisoner's Dilemma: Defectors—or to employ the terminology here, players who do not play HJV—need to be punished if they are not to multiply. An advantage of Four Temperaments players in that regard compared to utility robot, "play HJV/always cooperate" players is that the hypothesized programming for Four Temperaments players includes a Choleric humor, which is very well suited for punishing the uncooperative.

Melancholy/Partial Disharmony Games

In Partial Disharmony, aggressive and passive-aggressive behaviors pay off. Aggression is met with acquiescence. For an organizational story, consider Managerialism, in which bureaucratic, managerialist modes of operation give a government agency, or a for-profit firm, an edge over a player that adopts a Leadership/Followership orientation, even though both entities would do better with Leadership/Followership. A biological Partial Disharmony story could be called Exploitation: Proto-eukaryotes with a phagotrophic, predatory way of life gain an edge over their peaceful, prokaryotic neighbors.[7]

Though the two games are not identical, the solution concept for Partial Disharmony turns out to be the same as for the one just reviewed for pure Disharmony. In Partial Disharmony, just as in pure Disharmony, the Sanguine active subself is the first hero, playing Highest Joint Value regardless. The Melancholy reacting self is the second hero. Finally, the punitive side of the Choleric active subself polices the Phlegmatic reacting subself, and deters incursions by pure egoists.

Phlegmatic/Imperfect Harmony Games with Equal Outcomes

On the face of it, these are the most benevolent games, after Harmony. The reason is that in Equal Imperfect Harmony, the best response to another player playing a Highest Joint Value strategy is to play Highest Joint Value oneself. There is thus a good win–win outcome available for both players, without the tricky issue of inequality in reward that is the defining feature of Unequal Imperfect Harmony. On the other hand, there is an important complication that makes both Equal and Unequal Imperfect Harmony difficult. If the other does not play Highest Joint Value, one's best response is also to deviate. That means that both forms of Imperfect Harmony have two equilibria—a better one in which both players play HJV, and a worse one in which neither do.

For an organizational version of Equal Imperfect Harmony, recall The Last Shall Be First, in which a weak firm sets an example for other stronger firms, and in doing so helps create Highest Joint Value. For an individual version of Equal Imperfect Harmony, recall Follow the Rule, in which one mother who internalizes a rule of only looking for high-value fruit impels the other mother to also do so. For a biological story of Equal Imperfect Harmony, I will rely on the philosopher Brian Skyrms.[8] At a broad level, Skyrms asserts, in my judgment correctly, that the Stag Hunt—Equal Imperfect Harmony in my terminology—with its high-value and low-value equilibria is a better model than the Prisoner's Dilemma—or Disharmony—with its one, low-value, equilibrium, for analyzing how a low-value "state of nature" equilibrium gives way, or not, to a high-value "social contract" equilibrium. At a concrete level, Skyrms considers simulations showing that interactions involving neighbors can be more effective in moving players to higher value equilibria than interactions with random partners, and also considers biological examples similar to those I have relied on here to illustrate social games. He describes how the bacterium *Myxococcus xanthus* solves the Stag Hunt by aggregating into mounds when food is scarce, while living as individuals under conditions of abundance.

Now, let's analyze how *Myxococcus*, the mothers, and the firms solve Equal Imperfect Harmony. Here, the Melancholy reacting subself is the hero. Neither the Sanguine nor the Choleric active subselves have a Dominant strategy. Reactive Shame, on the other hand, does. It sends you no signal if the other deviates from HJV, but it wants you to play HJV if the other player does, which there is some chance it will do. You thus play HJV, and so does the other. Melancholy with a tincture of self-punishing Choler solves Equal Partial Harmony in our analysis here, just as it did in the Follow the Rule story of the mothers, in which internalized shame—guilt—leads one mother to follow a rule of only picking fruit, and the other mother to avoid social shame by following in her footsteps.

Harmony

In Harmony, the players' interests—of whatever kind, including sympathetic as well as selfish ones—are nicely aligned. As we have stressed, Harmony Games, much as they are less difficult than other social games, have their own considerable complexities and subtleties. The relevant complexity here is whether a Harmony Game that is readily solved by two purely self-concerned (or altruistic, ashamed, etc.) players is still readily soluble when the players are divided into the subselves we have hypothesized. The question: Does the division of the self into quadrants disrupt the Harmony that prevails with simple egoism, or other simple emotions and intuitions?

For an organizational version of Harmony, recall Respect/Love, in which participants in the social order are joined by respect, or at moments love, for one another's contributions. For an individual account of Harmony, we have many versions to choose from. Recall the quick alternation in the beginning of *The Hunger Games* between Katniss's Sanguine Harmony with her sleeping sister and mother, her Melancholy Harmony with them as they awaken to another sad day, her Choleric Harmony with her cat, and her Phlegmatic Harmony with her knife, boots, and other material objects. Or, if one prefers, recall the slower-paced System 2-rich alternation at the beginning of *The Tragic Muse* from the Phlegmatic Harmony of the energetic, self-assured British travelers in Paris, to the Melancholy Harmony in the family's shared downcast expressions and private anxieties, to the Choleric Harmony in the mother's proud and defensive relation to the world, to the Sanguine Harmony in the son's pleasant demeanor and the sympathy it engenders.

Finally, for a biological story of Harmony, we can use an account of how solo yeast cells combine into a multicellular snowflake, which was previously noted in another context.[9] For each cell, combining with others is a Harmony game. It's slightly too bad for the others if a "crazy" or "irrational" cell acts in a Disharmonious way by not joining the multicellular whole, since the whole will then be slightly less big and likely to survive. But regardless of what any other cell does, you "want" to join the snowflake.

Does the Harmony that prevails with simply programmed, unitary players still work when we have complex players, programmed with all four temperaments? The quick answer is that it does. You as Player 1 can rely on the Sanguine subself in Player 2 to play Highest Joint Value. Doing so is better for it regardless of what you do. "Knowing" that about the other player, both your Phlegmatic and Phlegmatic reactive subselves converge on playing Highest Joint Value. Harmony with players with simple egoism (or any other simple emotion) remains Harmony with players with four temperaments, even though the road to the solution becomes slightly trickier.

Technical and Ethical Dimensions of the Four Temperaments/ Four Elements Approach to Modeling Nature

On technique: I stress that the hypothesis, or conjecture, advanced here about how subselves of organic and inorganic selves can solve social games is only that. It would be of great interest to see how that hypothesis stands up to logical and empirical inquiries of the kind that modern natural scientists and social scientists excel in performing.

On ethics: A central reason game theory is controversial in ethics is that its key terms, though usually used as though they are factual and morally neutral, are subject to different, morally fraught, interpretations.[10] In the standard, positive sense in which game theory is generally taught, the players of a game are individuals whose outcomes, or payoffs, from different actions, or strategies, are dependent upon the strategies played by the other player. In a normative treatment of game theory, "game," "players," "individuals," "outcomes," "payoffs," "actions," and "strategies" are all subject to inquiry. So, too, is the central solution concept of Dominance. So as well is the idea of HJV, which is not widely used in standard game theory, although the concept (though not the name) appears in work by Schelling and others, but which is a key part of the analysis in this chapter.

In normative game theory, the same duality that we discussed earlier in "game" between egoism and competition on one side, and social feeling and shame on the other, inheres in other key terms. A "player" may be understood as a thing on its own, as an individual, separate gene or a Norway rat, subject to the logic of selfishness and the temptation to hurt competitors, and to the difficult dilemmas associated with that logic. But, as we have seen, any given player, any given individual, may also be understood as a part of a larger whole, like cellular mitochondria or a member of a group in a hypersocial species like humans or termites, subject to the logic of group-oriented altruism and the pain of punishment administered by oneself or the group for deviating from that logic, and to the different but also very real dilemmas of that logic. Likewise, "outcomes" or "payoffs" may be understood in terms of monetary or other rewards to a self-interested or competitive player. But they may equally be understood in terms of what is rewarding and hurtful to an altruistic or ashamed player. The gratification felt by others, or felt by oneself for helping another, the pain felt by others, or felt by oneself at the pain of another—these subjective outcomes or payoffs, as well as their objective correlates, are very much part of a sensibly inclusive normative game theory, and for that matter of positive game theory rightly understood. Including them makes it very difficult, indeed well nigh impossible, to figure out what a player's payoffs in a given situation actually are—but if the argument of this book is right, it is exactly that fundamental ambiguity as to the game we are playing in any given situation that allows self-aware human beings to focus on creating Harmony.

Likewise, "actions," if we prefer the more inclusive, less consciousness-implying term, and "strategies," if we prefer the term that gets at the idea of interdependence in the players' outcomes more clearly, both have a dual significance. Both terms may refer to behavior, whether calculated or unthinking, that implements egoistic or competitive programming in the player's evolutionary code. But both "actions" and "strategies" may equally refer to behavior that implements altruistic or socially conformist programming in the player. The same point applies to the central solution concept in standard game theory, in both its classical and evolutionary versions, of "dominance," or a "dominant strategy." The idea is that players will play a strategy that gives them better outcomes or payoffs, no matter what the other player does,[11] whether they do so through the classical theory's logic of calculation, or through the blind logic of evolution.

The same point about ethical and practical ambiguity applies to HJV, which is both a central idea for an ethically oriented, socially oriented, normative approach to game theory, and itself a deeply fraught, tension-laden notion. The basic idea of HJV is in one sense a simple one. Just as there are outcomes that are better and worse from the point of view of individual players, whether those players are egoistic, altruistic, or something else, there are outcomes that are better for the two players together, and for the social whole of which both players are a part. If we are to make progress in looking at game theory through lenses that are ethical, rigorous, and logical, though not mathematical, I believe that we need a notion of higher and lower shared value, social value, social welfare, or, as I have chosen to call it here, joint value. Without such a notion, the project lacks the constraint that systems of thought need to flourish. At the same time, the notion introduces a formidable array of ethical concerns, tensions, and conundrums.

A central tension in the notion of HJV is the familiar dichotomy of egoism and altruism: On the face of it, the concept of HJV seems to favor the social over the individual, just as the concept of higher and lower individual payoffs seems to favor the individual over the social. But just as the standard notion always entails a mirror world in which the individual is a part, the notion of HJV always entails the mirror world in which the whole is an individual; any given whole, from a body to a society to a universe, may be seen not only as the end, the *summum bonum*, of value but as an individual, existing and advancing itself, or not, alongside other bodies, societies, and universes.

The tensions in the idea of HJV also inhere at a more fine-grained level. To note the one that seems the most powerful and pertinent in relation to ethics: Should HJV be assessed simply through measuring utility, independent of the person or thing that experiences them? That part of Bentham's approach is consonant with the logic of wholeness and social totality. Some of us would follow Bentham in carrying our desire for logical consistency the whole way into a normative ethics of the social: We want the entity to flourish—and hang the individual parts, the mitochondria and the cells, that make up the entity! But many of us, most of us I suspect, will believe that we need a notion of HJV that is not Bentham's notion of simply adding up utils,[12] that cares about the parts of the whole at least to some degree, if only because any whole itself is part of another whole. But how much to care about the part, and in what way? Should we adopt an approach under which no part can be treated as a mere means to collective flourishing? How would that work? Do we adhere to a radical anti-consequentialism a la Kant,[13] in which the moral law is divorced from results, and in which our ethical interest in game theory, real though it may be, is focused solely on procedures and rules, such as a unanimity requirement for constitutional ground rules[14] that operationalizes the "do not use individuals as a means to an end" principle? Or do we follow Aristotle and virtue ethics in caring to a degree, though not with Bentham's monomaniacal intensity, about consequences as ethically significant, as I believe we should (though I will not attempt to demonstrate that fundamental proposition here)?

Some of us who have a real, if imprecise degree of ethical concern about consequences will be attracted to Rawls's answer to the conundrum of HJV: a class of the worst-off, however defined, must benefit from the inequalities (in primary goods, not utility, in his formulation) that benefit the whole.[15] Others of us will find his answer unattractive for one reason or another.[16] My own sense is that it makes sense to look for alternatives to Rawls's effort to solve the problem of social value and individuality. Rawls was 1971. We know much more about evolution and its logic then we did then. Theory needs to move on. But even those of us who find his vision normatively flawed from one perspective or another should, I believe, reflect on, and be guided by, the way in which Rawls conscientiously and brilliantly grappled with some of the deep riddles with what I have termed HJV.

To circle back to the theme of ethical duality and plurality within the basic concepts of game theory: Not the least attraction, and perhaps the greatest one, of game theory for ethicists inheres in the way that, far from eliding ethical tensions, it surfaces and heightens them. The foregoing discussion of the ethical connotations of basic game-theoretic terms may be taken as adopting a stance of moral superiority toward standard mathematical game theory on behalf of a normatively oriented, ordinary language approach: "By understanding the repressed normative tensions of standard game theory, we can develop a new theory that transcends such tensions!" As strongly as possible, I want to disclaim and reject any such position. Ordinary language does, I believe, have an advantage over mathematics in regard to surfacing and analyzing normative tensions. But in doing so, the ordinary language, humanist approach to game theory advanced in this book multiplies such tensions, rather than dispelling them.

Scientific and Ethical Lenses

Viewed as science, the basic Four Temperaments hypothesis, or conjecture, about subselves within organisms and inorganic matter can be related to a research program. The idea is that we can make progress by exploring ideas about the nature of human beings, all other organisms, organizations, and all forms of matter that are simple, and hence possible to define and analyze with reasonable precision. The basic idea that has been explored in most of the book is that humans are multiselved creatures. In this speculative chapter, the idea has been extended from humans to all players of games. In the formulation advanced here, players achieve HJV by valuing their personal utility, valuing the utility of others, disvaluing their personal utility, and disvaluing the utility of others. Or, to put the phrasing in a reverse form that highlights elements of contradiction: Players programmed to achieve HJV both disvalue their personal disutility and value their personal utility. Further, they both want other players to do well and want them not to do better than they do. They are, to use a human term, deeply ambivalent. One might say they are self-contradictory, because they have opposing, self-canceling preferences.

Now, let's switch to the terms used in ordinary speech, and by ethicists rather than by economists and other social scientists. The basic conjecture about human nature, and the nature of all other things, begins in a similar vein: We are divided creatures, with four main parts, or quadrants. In one quadrant, we are self-loving and self-concerned. In a second, we are empathetic, sympathetic, and serene. In a third, we are proud and indignant. In a fourth, we are ashamed and submissive.

In an ordinary language, normative ethical mode, the sense of internal contradiction among human desires that we have seen emerge out of the scientific version of the conjecture about the nature of all things does not arise. Ethics operates as a universal ranking device that allows us to avoid the apparent contradictions in our nature, and the nature of all other things, that appear when our desires are described in amoral terms. It allows us to make sense of our apparently disparate quadrants: We should, we intuit, feel proud when it is good to do so—as it sometimes is—ashamed when it is good to do so—as it also sometimes is—wrapped up in our feelings when that is good, detached from our feelings when that is good, solicitous of other people's welfare when it is good to be that way, and indifferent to their welfare, or outright opposed to it, when it is good to be that way. Having one's emotions ordered according to such a logic of ethical appropriateness is a hallmark of a person of good character. A person with emotions so arrayed is not a deluded soul who is repressing unresolved contradictions, but rather a person in full who is drawing on a wide repertoire of emotions to be exactly what a morally worthy human being should be.

What haunts the ethical version of the conjecture made here about four-part human nature is not the sense of incoherence and internal contradiction that haunts the scientist who has purged moral ranking in the course of performing her Phlegmatic scientific duty. Rather our problem as ethicists, who are open to the Choleric as part of our role, is doubt about the validity of one's intuitive sense about worthy human character. "Sez who?!" we ordinary language speakers, story-tellers, and ethicists wonder—or should wonder—about our assertions about human worthiness.

Finally, when we extend our ambit from humans to all things, as we have done, we are faced with the issue of purpose in all things that lies at the core of this chapter. Are we to regard animals, organizations, and matter as possessing an ethical character, if we believe that they have an ability to solve social games through positive and negative, active and reactive Four Temperaments programming that they share with us? Given this book's definition of ethics as the solving of social games, an affirmative answer is called for, I believe. I acknowledge the conundrums—but I would submit that one may reasonably feel a sense of fellow-feeling, of shared ethical being, not only with a squirrel or a tree, but also with a rock, a screwdriver, or a corporation.[17]

Classical Game Theory Revisited: A Story

The central point I have made in this chapter about the unresolved, unresolvable duality between parts and wholes in evolutionary game theory also applies to classical theory, with its focus on calculating individual human beings. We can flip the usual classical stories of freely choosing individuals into stories of individuals who are induced, programmed, or required by a social whole to act automatically, intuitively, or emotionally for the benefit of that whole. So, for example, instead of seeing the classic ratiocinators Holmes and Moriarty as choosers, playing the zero-sum game analyzed by von Neumann and Morgenstern in their pioneering work on classical theory,[18] one can envision them as programmed players of a social game. In that game, they are collaborating, perhaps to the greater benefit of one man than the other man, but to their mutual benefit and that of a whole—the mind of their creator Arthur Conan Doyle, it could be, or the consciousness of readers to whom Conan Doyle appeals, or human society writ large—of which both men are but dependent parts.

Suppose Conan Doyle, perhaps inspired by his own struggle to Harmonize both with his reading public, who wanted Holmes alive, and with his own desire to move on as a writer and to kill Holmes off, had published a companion story to "The Final Problem," written from the viewpoint of the master criminal Professor Moriarty and his consort Eugenie, rather than from that of Holmes and Watson. Doyle could have flipped the scene in "The Final Problem" in which Holmes outguesses Moriarty, and he and Watson watch Moriarty's train steaming ahead. In the flipped story, we can imagine Moriarty pacing restlessly in his plushly appointed train car, the great dome of his forehead glistening, and reflecting to himself and Eugenie:

> Why did I change my mind and give the order for the train to steam on past Canterbury? It is a nice puzzle! You do not know this—but before I gave the order I had seen Holmes there at the station, clear as day through my binoculars, smug as ever, not seeing me. The fool! A confrontation with him might have gone either way, but the odds of course favored me and my henchmen.
>
> My answer is . . . ah, the true answer is that I do not know, Eugenie. I wanted to play backgammon and read Zola with you. But there is more to it, I am sure. Part of me needs Holmes, and part of him needs me. How to explain that need? Let me elucidate. When I was a boy, and not yet the bad, perhaps mad, man that I have become, I always rooted for heroes. At the same time, contrary as I was, I always wondered why the villains in the stories I read and saw on Drury Lane did not simply kill the heroes when they had them in their clutches. Now, as I reflect on what Mr. Darwin and Mr. Spencer have taught us, it seems to me that it is valuable for those of us who are the villains in our great social drama to have a reluctance to finish off those like Mr. Holmes who are the heroes. And the same is true for him, though my theoretical calculations show that his reluctance to kill me will be less than mine to kill him.
>
> Yes, Holmes and I chase one another through the serene English countryside with pistols at our sides, and accomplices with their own pistols. And yes, in some cases the game we play will demand that one or both of us will die. After all, we cannot let the great public know that we are, in a sense, in league. We must go along with their image of the two of us as locked in mortal, eternal conflict. But we must also go along with the greater game that the whole social organism is playing with us. Who comes out the better in the game? Is it me, with my wealth, and with you, Eugenie—thank God I do not have to spend my days with Watson! Or is it Holmes, with the esteem of the world, and his ever-present warm glow of superiority?
>
> I pondered that question for years, my darling. Now I know the answer. It is neither of us! We both lose! I may die soon, and our love may be lost forever. And so too for Holmes. It is indeed a most dangerous game that Holmes and I play. We are twin Leaders, he and I, who both lose relative to the great inert mass of human cells who Follow and risk not. The player that comes out best in the game we play, I am sure, is the great social beast that sits comfortably

in its millions of homes in its corsets and its smoking jackets, reading its newspapers. That beast feeds on both of us!

Here is how it works: Human morality if it is not simply to be a matter of rival groups tearing one another apart needs to have a clear focus, a negative model of what all decent people of whatever group are against. I am that model! I need to exist so the great social beast can know what is bad, and can know that it is good by comparison. And yes, the people need a model of the good, of reason and virtue conjoined, just as old Socrates dreamed. There is no plausible saint these days—but we do believe in science, and Holmes with his brilliance and his goodness is our Socrates.

But that is not the end of it! There is a great beast of the universe that pulls the strings of the newspaper-reading beast that pulls the strings of Mr. Holmes and me—the beast of Evolution!

Eugenie looks bored, and Moriarty realizes it is time to end his soliloquy.

So. So. Do I hate the great beast that pulls every string? No! Not at all! It is the same beast that makes you beautiful, and that makes our games joyous. But do not worry, my love! I do not worship the beast. There is only one I love, and you know who she is.

Let's play, shall we?

* * *

My closing suggestion in this speculative part of the book: It is fine for Professor Moriarty as an ethically unhealthy man not to worship the great beast that has made him what he is. But it would also be fine—indeed, it would be a very good thing—for the great mass of us who are ethically healthy, more or less, to revere the ethical logic of Nature.

Let us, as we are willing and able, cultivate ways of feeling and thinking that allow us to experience awe, not only at the beauty of the starry skies, but also at the truth and the goodness conjoined with beauty that shines forth from the stars, from human faces, from every creature, and from every thing, including every created thing, from backhoes to business corporations. In addition to the Phlegmatic, Melancholy, and Choleric forms of reason at which we moderns excel, let us open the door wide to the Sanguine form of reason that our ancestors knew, and that at some moments, though never always, we can know more fully once again. Let us see the telos in Nature. Let us move, as we are willing and able, from Respect to Love (Figure 4.1).

> **The Four Temperaments/Elements as Solutions to the Four Games**
>
> | Games | Active/Yang—
Dominant Strategy ! | Reactive/Yin—
No Dominant Strategy ... |
> |---|---|---|
> | Positive | **Harmony Games**
Solution:
The Sanguine in Nature | **Imperfect Harmony Games**
Solution:
The Phlegmatic in Nature |
> | Negative | **Disharmony Games**
Solution:
The Choleric in Nature | **Partial Disharmony Games**
Solution:
The Melancholy in Nature |
>
> It is reasonable to believe that the positive-negative, active-reactive arrangement of moral emotions into temperamental quadrants helps solve social games in human beings. It is also reasonable to conjecture that the same positive-negative and active-reactive arrangement, transmuted from the psychological level to the biological, chemical, and physical levels, helps solves social games involving non-human animals, organisms other than animals, inorganic matter, and created entities such as firms.

Figure 4.1 Telos

> *Summary*
>
> In this speculative chapter, I suggest that the logic of the Four Temperaments can be universalized from human beings to other animals, other organisms, other forms of matter, and to created entities, such as corporations. I suggest that in all these cases there is a parallel logic, and ethic, of the part, and a similarly parallel logic, and ethic, of the whole. Evolution may be understood as the working out of that fused logic and ethic.
>
> The Four Temperaments, understood not simply as human emotions, or intuitions, but as positive and negative impetuses to action—the Sanguine and the Choleric—and as positive and negative impetuses to reaction—the Phlegmatic and the Melancholy—may be the underlying basis for the solution of the four major kinds of social games. Whether that turns out to be the case or not, or whether the question turns out to be answerable or not, the colorable possibility of explaining the universe of all things interacting with other things in terms of a basically benevolent ethical logic is important. It allows us to bring back telos, a sense of purpose in everything, which for so long so many have considered hopelessly lost.

Exercises

1. Compare the sense of oneness with everything and universal goodness that Levin experiences at the end of *Anna Karenina* with the sense of separation from everything that Mearsault experiences in *The Stranger*. Discuss, or reflect on, whether you believe, or feel, Mearsault's Melancholy sense of meaningless is more grounded in scientific truth than Levin's ecstatic, Sanguine sense of oneness; further discuss, or reflect on, whether the case made here for the ethical nature of evolved things changes that belief, or feeling, or not.
2. Search for, discuss, and/or reflect on, online games that simulate cultural evolution. Are the perspectives of these games parallel to the optimistic perspective on social games here, or not?
3. (a) Read and discuss, or reflect on, the Harmony-oriented account by Lynn Margulis on how complex eukaryotic cells evolved and the Disharmony-oriented account by Gaspar Jekely of the same process; (b) Read about, watch videos on, or, if it is possible, do experiments on the evolution of snowflake yeast with the materials provided by Will Ratcliff; (c) Read and discuss, or reflect on, the interview with Cassandra Extavour on the use of normative terms like cooperation in biology.
4. (a) Research the companies that have lasted over time on the Dow Jones Index compared to those that have dropped off the index; do those results support the idea that companies with a more ethical nature do better in solving social games and in flourishing over time, or not?; (b) Discuss, or reflect on, experiences of yours with companies you have dealt with as an employee, customer, supplier, or another capacity that either support, or conflict with, the idea that more ethical companies are better at solving social games and are more likely to succeed than less ethical ones.

PART II

Business Ethics

CHAPTER 5

Critical Business Ethics

My aim in this chapter is to provide some practical ideas on how the Four Temperaments approach to game theory that this book is devoted to promoting can be related to teaching, and also to research and practice. Although the material in the chapter can stand on its own, it can also be related to a potential critical business ethics school, and to analogous groups in other disciplines, that bring together people who share an emotional commitment both to scientific truth and to fictionalizing, fabulizing story-telling, and who respect and embrace both a critical, debunking spirit and a Sanguine, accepting one.

In the Preface, I told the story of how I moved from my Thomas Schelling-inspired enthusiasm about game theory to a disillusionment about the field. Influenced by work by Duncan Kennedy and others, I became convinced that there was a profound gulf between the calculable rationality of means, as analyzed by game theory, and by economics more generally, and the elusive, but necessary, rationality of ends that was the domain of law, politics, and ethics.[1] I became committed to the position that the Prisoner's Dilemma and its logic could be appropriated equally by liberals, conservatives, centrists, radicals, Buddhists, and paranoids to advance their respective passions. Logic was not a liberal, as I had once hoped, nor a conservative, but a vehicle for hire, available to take one to the end of any line one chose to ride—or so I believed, and so I tried to show in a series of debunking articles.[2]

As this book testifies, I have changed back to becoming an enthusiast for game theory, as long as the field is defined, as I have tried to define it here, to include a humanistic, moral emotions wing to set alongside its classical, scientific wing. At the same time, I still find myself deeply attracted to thinkers, like Hume with his guillotine between is and ought, Max Weber with his divide between fact and value, and Professor Kennedy with his division between logic and law,[3] who have a powerful self-denying, self-punishing, ascetic element that revolts against the towers of reason that they are exceptionally skilled in erecting. The basic spirit of the present work is Pharrell's "Happy," not Bob Dylan's "Tears of Rage"[4]—but we very much need both kinds of song.

The Litmus Test

A commitment to the moral emotions take on game theory advanced in this book can be combined with all kinds of temperaments. That said, I believe that commitment works best for people with a strong temperamental attraction to both science and technology and to literature, or, more broadly, to art. Further: I believe it works best for people who are powerfully attracted to both the intuitive System 1 and the reasoning System 2 sides of both domains. If one reveres Henry James, but is left cold by Suzanne Collins, fine—but unless one feels the pull of popular culture, with its consonance with our predominantly System 1 selves, one will, I believe, be outside the circle of appreciation for the approach advanced here, even if one finds it plausible intellectually. It may be John Lennon, Beyonce, Scarlett Johanssen, or less renowned popular artists who light your fire—but someone must. Similarly, fine as it is to be unmoved by Henry James, unless one can feel the lure of high art that draws strongly on our System 2 capabilities, one is similarly outside the circle.

The same proposition applies to science. One may not truly understand, say, Godel's Theorem—despite puzzling over the applicable parts of *Godel, Escher, Bach*[5] for hours as a young man, I never did—or, for that matter, care about it. But one must be compelled by the magic of high science, and by the abstract, calculative logic that is its indispensable accompaniment. On the popular, technological side, one may carry around an old flip phone, or no phone at all, have never played a video game, and be indifferent to the latest Apple devices and apps. But to emotionally get the approach advocated here, to feel it in one's bones, one needs to feel excitement at popular science, that is, technology. A major part of what is advocated here is Phlegmatic Harmony between people and things; technology is a central incarnation of Phlegmatic Harmony.

In the union of the four parts lies the possibility of a critical movement of teachers, scholars, and practitioners. In what manner would such a movement be critical? The basic answer is that, in one way or another, varying according to their circumstances, the people drawn to such a movement would be impelled by a desire to bring together, and to balance, all too often sundered, all too often unbalanced elements of both high and low science, and of both high and low art, in their speech, their writing, their other actions, and their environments. In welcoming people of all political persuasions, such a critical movement would be, in one sense, more inclusive than traditional critical movements, such as critical management studies and Critical Legal Studies.[6] But in expecting those committed to it to bend the knee to both science and art, and to both high culture and popular culture, it would be less inclusive. Opening the door to Sanguine reason should be for everyone—but a movement devoted to doing so will not be. The model of game theory advanced here is for everyone—but the utility of the model to inspire will, and should, vary.

Rhetorical Experimentation

Academic business ethics is a field sundered between a normative and an empirical side.[7] If we matter, it is partly because of our canary in the coal mine status as a split field, with a more intense version of a divide that pervades modern intellectual culture. In addition, business ethics has a wide divide between our academic side and our practical side. Tom Donaldson on the theoretical side and Dov Seidman on the practitioner side[8] are both very good in their respective ways, but they are very different from each other. Can we Harmonize as a field, in our writing as well as in other ways, among our social scientific, our normative, and our practical sides? Can we bring high art and popular art, along with high and popular science, into business ethics? Can we tell new stories that conform to the elaborate, tacit protocols of Harmony that govern whether a text is in right relationship with a reader?

On rhetoric, I have a simple idea to suggest. Let us say the goal of providing an alternative to the typical corporate, one-voice style of jointly authored academic and practitioner papers—which is its own worthy, if Melancholy, form of Harmony—is a good one. Then let's experiment as academics and practitioners by playing a Sanguine Harmony game in which we write in our own voices, and Harmonize in the quicksilver moods and thoughts of our coauthors, who in turn will Harmonize with us in an ever-shifting succession of games.[9]

A conversational, dialogical form of jointly authored writing that fuses academic and practitioner styles—critical business ethics, in one version of that genre—may or may not be successful in Harmonizing with readers. But it is, I believe, an experiment worth trying to make work. "No ideas but in things," New Jersey poet William Carlos Williams said.[10] No ideas but in people and their diverse, fleeting intuitions and emotions, a critical business ethics credo might read.

Bringing Psychology and Economics Together

For many decades now, psychologically oriented management professors have been excellent at generating case studies, empirical results, and theories about that support the value of optimistic, collaborative management.[11] As one whose father was a management professor who taught organizational behavior and whose mother was a psychologist, I grew up with that perspective on management. I continue to believe in it. At the same time, I respect the concerns of those with an "incentives matter" bent, to whom management theories and empiricism often seem too good to be true.[12] I want to suggest here that the perspective on game theory advanced in this book can help us think through the tension between the aspirational "yes we can" and the skeptical "it's all about incentives" sides within business schools, within business, and within ourselves.

In this section, I want to focus on a very nice, eyebrow-raising empirical result, and then move from that particular study to more general issues. I heard the study presented a few years ago at a supply chain management conference by a young professor, Brad Staats. I was there as chair of recruiting for my department, with the aim of meeting supply chain PhD students who did field and lab experiments that tested an economic, rational choice model of human behavior against psychological, moral emotions models of behavior. Brad's study was quite different, and it both impressed and disconcerted me.

When I searched online to write this section, I quickly located the published article connected to the presentation: "Breaking Them In or Revealing Their Best? Reframing Socialization around Newcomer Self-Expression," coauthored by Cable, Gino, and Staats.[13] As I read the abstract, the study came back to me, in the "Aah, yes!" rush of awakened memory in the Harmony game that our past and present selves play.

The authors had worked with Wipro, an Indian business process outsourcing company worried about high quit rates, to test the efficacy of three styles of onboarding training that employees received on the first day of work. The first kind of training focused on expressing one's authentic self, while the second focused on instilling organizational pride, and the third, the one Wipro had traditionally used, focused on instilling skills. The results of the study were powerful: New employees in randomly selected call centers who received an authentic-self welcome to the company were far less likely to quit, and did better on numerical performance indicators, than new employees who received training focused on loyalty or skills. In addition to their field work in India, the authors had conducted lab experiments in the United States comparing the effects of training oriented toward self-expression and training oriented toward collective goals on whether American university students returned on another day to do more work, and how good their work was. With the American students, they had found the same results as with the Indian workers—the "be yourself!" training worked better. And even if the real-world significance of college students returning voluntarily on later days to do more work and better work in a lab wasn't as impressive as the call center results, it was still impressive in its own right.

Fascinating as they were, the results in the Indian field experiment and the American lab experiment in favor of a "be yourself!" approach to managing did not provide a mechanism to explain why the empowerment approach would work better than the conformity and skills approaches. Nor did the results provide a mechanism to explain why any of the managerial approaches to first-day training would cut through the gritty, often much less than pleasant day-to-day realities of call center work to make a difference in how people performed their jobs and whether they stayed on the job. I might agree—I did in fact—with the authors that people yearn for authenticity, as well as for more prosaic things, around the world and in all kinds of places, in tough jobs as workers in call centers and factories, not only in plush jobs as professors and executives. But the skeptical, economically oriented side of me wondered: How are these yearnings to be modeled, and how are they sustainable over time?

I think the skeptical question is an important and a difficult one at the level of management theory and practice, as well as at the level of philosophy. I want to try here to connect work in call centers, and the academic effort to understand it, to the human drive to interpret ambiguous social reality in Harmony terms that has been hypothesized here.

I would suggest that the different modes of training at the call centers studied by the authors can all be understood as efforts to advance particular styles of Harmonization in the workplace. Under the perspective advanced here, all of them could be described as doomed to succeed, since people in social groups will create Harmony games, in one way or another. But not all Harmony games are equally productive from an organization's perspective, whatever their value may be for the individuals who play them. And not all Harmony games that work well for an organization are sustainable from the perspective of the individuals who play them.

From the perspective advanced here, the key question about the study and the real world is whether the result—that emphasizing self-expression is a more productive style of management—has a plausible Harmony grounding. Why is a self-expression version of a Harmony game sustainable in a firm? Why might it, perhaps counterintuitively, be more sustainable than a corporate loyalty version of Harmony, or other versions?

It is not hard to envision how a loyalty, following social norms version of a Harmony game could be sustainable for a firm like Wipro: Employees who do not see life in the call center in Harmony terms can be stigmatized as disloyal, while employees in the middle can be disciplined by their own shame, as well as by their egoistic desires not to be punished, and to be rewarded for high commitment and high performance.

By contrast, the path to a Sanguine self-expression version of a Harmony game being sustainable for Wipro and for firms in general may seem less intuitively clear, or outright murky. With a culture that values individual authenticity, can one punish disloyal "take this job and shove it" sentiments, as opposed to punishing only poor results? What does one say to a call center employee who believes in the self-expression message, and says to a manager, "I don't feel ashamed at all about not caring about making a rupee for Wipro!" Given its apparently limited and unclear powers to summon the big guns of righteousness and shame against employee recalcitrance, how can a culture of authentic self-expression in the workplace, whatever its merits for elite workers like university professors, be sustainable in the tough work environment of a call center or a factory?

The answer to the sustainability question I would suggest runs along the following lines: the optimists are right, in that an individualistic, authenticity-oriented approach to Harmony games can work very well over time, and for that matter can do so at a much broader level than an individual firm like a university or a call center. At the same time, the hardheaded critics who insist that there is no such thing as a free lunch are also right.

If a Sanguine, self-expression, individual empowerment version of Harmony outproduces other versions, it is not, under the partly aspirational, partly skeptical view advanced here, because we have created a new world that transcends the old world and its dark-side emotions of shame, guilt, and righteous anger. On the contrary: If a Sanguine, optimistic mode of Harmony is more productive for call centers, factories, schools, and other organizations in which we manage and lead, as I believe it generally is, it is partly because that approach works better than any other to make us our own policemen, who put the gun at our own heads—who are, in fact, the gun at our own heads—by deploying shame, guilt, anger, competitive pride, and other troubling, necessary moral emotions to make ourselves work and produce. Happy, empowering leadership fosters self-disciplined workers in the service of Phlegmatic, pragmatic business ethics. The leaders and the ethics are both positive. But we managers and workers make the ethics work with self-administered Melancholy and Choler.

The dark side is a central part of the truth. At the very same time, a sunny-side, aspirational explanation of the productivity of Sanguine leadership and followership is also correct, I believe. If the approach here is right, human beings are made to Harmonize, and to do so through rapidly shifting, mutable alignments of moral emotions, with one rapidly succeeding another. A Sanguine style of leadership that empowers people—and in doing so gives them the strength to hold the gun at their own heads—that does not impose a monolithic System 2 logic of argument, compliance, or calculation, and allows us to employ the full gamut of our moral emotions in our work, is likely better aligned with our basically, though not uniformly, good nature than any of the main alternative styles of management are.

How I Changed My Approach to Teaching Business Ethics

I used to practice law, and I began my academic career writing for law reviews, and teaching business law. When I started teaching business ethics some years later, the written work and the presentations were adapted from the model I used for my law classes. I had my ethics students write legal-style case analyses—on issues such as "Should Domino's Pizza suspend its 30-minute delivery guarantee?"—with boxed pros and cons, point–counterpoint arguments for both sides, followed by the student's own conclusion. The students did courtroom-style presentations with competing teams—on topics like "Resolved: Wal-Mart should pay its employees a living wage"—with other members of the class asking questions as judges, and then deliberating and voting on a verdict.

I still use legal-style case analyses and courtroom-style presentations as part of my business ethics teaching. I believe that one very important part of business ethics is the Harmony game of universal reason, and that in the cultural context of the contemporary United States, if not of all places and all times, competitive, Choleric Harmony debates with equally matched sides and then a judgment are one very good way to play that game. But some years ago, influenced by reading Jonathan Haidt and others, and by my own project of analyzing moral emotions and social games, I decided that I would no longer rely as exclusively as I had in the past on an argumentative, law-based approach to business ethics.

I now try to roughly match my adversarial, legal-style assignments with assignments that focus on more intuitive, emotion-oriented, social, and personal aspects of ethics. In addition to asking students to write point–counterpoint case analyses of what the right resolution to an ethical dilemma is, I now also ask them to write what I call ethical relations analyses,[14] in which a key person in the organization explains and tries to justify a decision to key groups such as investors and employees. In the appendices, I've included an ethical relations analysis I have used for the issue of Domino's

Pizza deciding whether to suspend its 30-minute delivery guarantee. As interested readers can see there, the ethical relations analysis uses a back-and-forth format, but in a way that has the student imagining the speaker, in this case Domino's founder Tom Monaghan,[15] making personal, often emotion-laden, statements to explain and justify two different hypothetical decisions he could make. The idea is to move students, and ourselves as teachers, away from the universalizing voice that comes naturally to us in writing in the direction of a more personal, group-oriented, relationship-oriented voice that comes naturally to us in face-to-face interactions.

I wrote the Domino's ethical relations template before working out the idea of Harmony and its manifestation in four temperaments, or ethics, that is developed in this book. Partly because I believe that imposing a Harmony framework is not the best way to present it to students, I haven't changed the template by restyling "ethical relations" as Harmony. I have found, though, that my template, and much of what my students write, can be understood in terms of repeated appeals to Phlegmatic Harmony—no surprise there, if that type is indeed central to management—and also to Sanguine Harmony, Melancholy Harmony, and Choleric Harmony; the notes in the Appendix include my recent classification of how Tom Monaghan's remarks to various stakeholders can be expressed in terms of the four types of Harmonizing.

For presentations, I now rely mostly on having students simulate meetings rather than trials, so that ethical concerns can get raised, and sometimes skated over, in a way that simulates the actual dynamics of human social life in organizations in a way the trial format does not. So, in a presentation on the WorldCom accounting fraud—more on that later in the chapter—and how it played out in the life of Betty Vinson, an accountant at the firm, and Scott Sullivan, the company's CFO and Betty's boss,[16] the students will simulate meetings. They will play out a meeting between WorldCom accountants about the plan to manipulate earnings, a phone conference of New York and Mississippi prosecutors discussing whether to prosecute Vinson and other lower-ranking employees, and a meeting of managers at KFC, where Vinson has applied for an accounting job after serving her sentence. Or, in a presentation on Wal-Mart, instead of debating the rights or wrongs of outsourcing and wage policies abstractly, presenters will take on roles as factory workers in China and managers in Arkansas. In doing so, I believe the class comes closer to evoking the distinctive Bentonville-Guangdong global supply chain management culture of cheap plastic office furniture, hula-dancing, bear-wrestling managers, and employees doing company cheers that Sam Walton, David Glass, and others created.[17]

I'm continuing to experiment with formats for written assignments and presentations, and expect I always will. I suspect that for many teachers, my recent efforts to orient my business ethics classes in a more psychological, social direction would have been the starting point. I'm still a work in progress after more than twenty years of teaching, and many other teachers—including my wife, who teaches high school English—have a more natural predilection for intuitive System 1 human social drama that helps them to create livelier Harmony games in their classrooms than I do. At the same time, I think that there is a positive connection between my intellectual turn toward psychologically oriented, moral emotions game theory and my teaching. I'm hopeful that students and teachers in business ethics, and other fields, can take the ethical relations approach I've used in my classes, and more broadly the perspective on humans as creators of Harmony games, as one starting point for their own experiments in advancing and creating knowledge.

Playing Games in Class

There is a natural tendency to think of games in terms of calculative, egoistic, and/or competitive logic. That is perfectly fine, I believe, if we can also keep in our mind's eye an equal and opposite thought of games as connected to the less precise, less calculative, but also highly potent and valuable, logics of human flourishing and human shame. That equal and opposite vision with respect to games is one way of expressing what the Harmony perspective of this book is about.

In my classes, I try to bring the idea of an equal and opposite vision down to earth by engaging my students in games that shake up an overly simple equation of them with egoistic or competitive calculation, but that at the same time acknowledge and respect the calculative spirit that is a highly potent and valuable part of business and business ethics, and of human nature. Inspired by Professor Schelling's example, I've experimented with many exercises and games over the years; since I teach business ethics, though, rather than game theory, in a semester I usually only do about as many games as we sometimes did with him in one class. In what follows, I present the five games I've devised that I believe have worked the best in fostering a pleasant "Aah!" or "Aha" sense, along with, I hope, cultivating in my students and in me a more nuanced sense of the relationship of games to ethics. I combine exposition of the themes that the games illustrate with descriptions of the games, and the drift of the results I've gotten. For interested readers, the actual exercises are in the appendices.

The Blame Game

I have an intuition that the biggest key to providing a more Sanguine cast to the ideological divides over business, and business ethics, lies in our understanding and appreciating, if not necessarily endorsing, the way human moral intuition works in assessing intention, responsibility, and blame. Joshua Knobe's finding some years ago that people believe a profit-motivated CEO who harms the environment as a side effect of the search for profit intended to do so, while one who helps the environment as a similar side effect does not intend to do so, is, I believe, one that all of us who teach business ethics and study it should be familiar with.[18] As shown in the matrix that begins the summary of this chapter, I've extended Knobe's work on how we evaluate calculative egoists, or part-oriented actors (such as his CEO), to include noncalculative egoists (such as young children), noncalculative pro-social, or whole-oriented, actors, and calculative whole-oriented actors.

I've gotten good results by dividing the class into different teams that take on different versions of the survey, and then report back to their classmates on their discussion.[19] To cite a recent result: In the team devoted to considering whether the profit-motivated CEO who harms the environment as a side effect of that pursuit does so intentionally, my undergrad class readily adopted the "yes, he does" perspective of Knobe's respondents. But my MBA team was split right down the middle; it delivered a 6–5 verdict in favor of yes, with the five dissenters articulating an argument against an unintended side effect being intentional. Is that a revealing indication, perhaps, of a split between outsiders and those of us who are in business, and who, in our moral interest, find it easy to come up with complex, exculpatory double-effect reasoning that is elusive to outsiders? Perhaps. It is interesting to note that, in the same class, another team answered "no, he does not" to the question of whether a CEO who holds protecting the environment above all other values, and therefore does not develop a product, with the result that the company is harmed, harms the company intentionally. I look forward to more informal explorations of this issue, and related topics, with my classes. More broadly, I believe that extensions of Knobe effect research by business ethicists, and for that matter by scholars in other fields, have considerable potential to advance the cause of Sanguine reason.

Ethical Focal Points

In Chapter One, I described how Schelling's idea of an intuitive human skill in coordinating on focal points can be applied to ethics.[20] The key empirical results: My students coordinated very easily on the one "good guy"—Abraham

Lincoln—juxtaposed with four bad guys. By contrast, they did not coordinate at all on the one "bad guy"—Charles Manson—juxtaposed with four good guys, splitting instead among the good figures.

In keeping with the protocols of empirical social science, there are plenty of good, critical, practical questions to ask about my survey and its results. What happens if a very famous bad guy—Hitler would be an obvious choice—is juxtaposed with the good figures, instead of the now obscure Manson? What happens if instead of the altruistic "help poor people" framing I've used, we use an egoistic "make the most money for yourself" framing, or for that matter a villainous, Snidely Whiplash-style "do the best job tying Nell to the tracks" framing? What happens if we switch the order of the responses, so that Lincoln and Gandhi (the narrow victor on the second question) are no longer choice (1) but choice (4), say? As those of us who specialize in, or, like me, engage in, empirical research always like to say, further research on these questions, and others as well, is in order.

As much as I strongly support rigorously technical empirical social science as one important genre of System 2 reasoning, the empiricism in this chapter is aimed at social scientists, and the rest of us, who are wearing our explorer's "let's think about this . . ." soft felt hats, not the construction worker's hardhats we wear when we are in the world of top-tier journal publication. With that in mind, I offer a few reflections on the empiricism, the potential theory to be developed, and the philosophical implications with respect to ethical focal points. On the empirical side, the key "Aha!" moment for me was a sense that there may well be a powerful and highly useful asymmetry between a strong intuitive human ability to coordinate on a good focal point and a lack of such ability with regard to a bad focal point. Schelling focal points, I think, have a moral logic attached to them. I would speculate that this asymmetry works very well at the intuitive System 1 level, but not so well at the level of System 2, where we find ourselves attracted to reasoning through unpleasant dilemmas. As modest evidence in that regard, I would cite a question on my survey that asks students to choose between Harmony, Disharmony, Partial Harmony, and Partial Disharmony games. Whether because I have couched the Disharmony choice in terms of the very famous Prisoner's Dilemma, or because our System 2 side likes tackling tough problems, or for other reasons, Disharmony/the Prisoner's Dilemma has narrowly beaten Harmony as the focal point game for my students so far.

At the level of theory development, I believe that the elaboration of testable propositions related to focal point logic and Harmony is important for the moral emotions approach to game theory advanced here. The intuitive, example-based argument made in the first chapter that human nature is a conspiracy to Harmonize rather than play unpleasant games might be right even if it cannot be turned into good empirical social science—but the chance that it is correct goes down in that case, and it goes up if the empirical project is a successful one. At the level of philosophy: In my teaching, I am reluctant to claim too broad a meaning for the focal point survey; the click of connection to the cosmic is better if it comes from within the student, rather than from me. At the same time, I acknowledge a strong underlying belief that ethics can be usefully understood, along with speech and writing, as a master technology by which humans coordinate effectively with human beings and the rest of our environment, and a hope that teachers skilled in the Sanguine Harmony game of lighting the sacred fire through alluding without imposing may be able to make use of that thought in their work.

Judgments of Fairness and Unfairness: Self-Interested or Harmonizing?

At the heart of game theory is the tension that exists in all social games except Harmony between the principle of Dominance—play the strategy that is best, regardless of the other's actions—and the principle of HJV—play a strategy that together with the other player will allow you to achieve the best outcome. Tension becomes outright opposition in what are conventionally called Prisoner's Dilemmas, Tragedies of the Commons, public goods games, social dilemmas, or free-rider games, and what are here collectively termed Disharmony games. In Disharmony, as previously noted, the HJV strategy is dominated, and thus worse for you—and the "you," as previously noted, may be Mother Teresa as much as a Wall Street trader.

The basic question I've looked into in the next exercise is whether students' judgments about the fairness or unfairness of playing a Dominant strategy in a Disharmony game/Prisoner's Dilemma are affected by what happens to them in such a game.[21] I've asked my students to imagine a game with a classmate whose identity they do not know with the following rules: If both play the HJV strategy, both get 5 virtual extra credit points. If both play the Dominant strategy, both get only 1 extra credit point. But if one plays Dominant and the other plays HJV, the Dominant player gets 6 points and the HJV player gets skunked with 0.[22]

Now, the key part: In the "pre" part of the exercise, I've asked my students to say whether they think playing that a Dominant strategy in the game is fair or unfair. They then play the virtual Disharmony game by picking either Dominant or HJV, and then learn whether they have gotten 6, 5, 1, or 0 "points." Then, in the "post" part, I ask them once again to say whether they think that playing Dominant is fair or unfair.

There are two major conjectures that I've been interested in testing about how students might change their answers from pre to post. One guess—the more intuitive one, I believe—is that people's judgments of whether playing Dominant is fair will shift based on whether it has worked well for them or not. So, for example, if you get 6 points by playing Dominant while the other gets 0 points by playing HJV, great—you're now more likely to evaluate playing Dominant as fair. A contrasting conjecture is based on the Harmonization perspective of this book. The basic idea here is that winners as well as the losers in the Disharmonious 6–0 splits will be more likely to see playing Dominant as unfair after the fact. On the other hand, players who agree in playing Dominant or Highest Joint Value, and get either 5–5 or 1–1, will be less inclined to see unfairness in the post- than in the precondition.

Most of my students stick with their "pre" judgments, so I don't have a lot to work with yet after a few iterations of this exercise. But in the early returns, Harmonization as defined here is beating self-interest. Five of the six students who have switched from judging Dominant as fair to judging it as unfair were in disagreeing 6–0 pairs, and all seven of the students who switched from "unfair" to "fair" judgments were in agreeing 5–5 or 1–1 pairs. Harmonization has thus worked 12 out of 13 times. By contrast, self-interest hasn't worked well as a predictor. Losers in 6–0 splits do indeed tend to move from judging Dominant play as fair to judging it as "UNFAIR!"—as one student wrote, and as both the self-interest and Harmonization explanations predict—but overall, the self-interest hypothesis has explained only 4 of the 13 shifts.

The same points noted in relation to ethical focal points apply here. The claim that moral judgments of X, and our actions with respect to X, are in significant part related to whether our personal experiences with X have been Harmonious, rather than personally successful or unsuccessful, is a claim that should be further studied and subjected to the rigors of empirical social science, with our hardhats firmly affixed. The claim should also be related to existing research relevant to the empirical development of a Harmony perspective, such as Robert Cialdini's experiment, in which California homes receiving messages about their neighbors saving energy did a much better job Harmonizing by reducing their own consumption than homes receiving messages about saving money and protecting the environment.[23] At the same time, here as with focal points and the other exercises, my sense is that the greatest potential value going forward likely lies not in formal empiricism, but in creating a set of accessible, evocative Harmony stories, like those I offered in the last chapter, that parallel the ones that Schelling and others created for game theory in its postwar dawn. Teachers in business ethics, game theory, and other fields can learn, improve, and pass on these stories.

The Ethical Wisdom of Crowds

In James Surowiecki's book some years back, two of his signature cases on the wisdom of crowds[24] involved the ability of groups of people collectively to come up with highly accurate guesses of the number of jelly beans in a jar—he cited a finance professor whose class of 56 students came up with an average guess of 871 for a jar with 850 beans, better than all but one of the individual guesses—and of the weight of an ox—an average guess of 1,197 pounds versus the actual 1,198 pounds in an early twentieth-century British contest in which 787 people entered. Is there a parallel in ethics, with people in groups having an uncanny ability to solve certain social games?

Based on a simple survey I recently rolled out with undergraduate and MBA business ethics classes, I think the answer may well be yes.[25] I gave my classes a hard—I thought—Partial Disharmony game in which I asked them all to pick a number from 1 through 6, with the following rules: If the class as a whole had an average under 4, everyone picking 4 or over would get 3 virtual points, while those picking 3 or under would get a good, but less good, 2 points. On the other hand, if the class had an average of 4 or more, everyone would get only 1 point. The students were not allowed to discuss their picks with one another.

There is a large body of behavioral game theory experiments from the 1950s on showing that two people do better in one-shot Disharmony and other tough games than traditional game theory predicts.[26] For example, players in one-shot, two-player Disharmony tend to reach HJV about half the time or a bit under, compared with the flat zero they achieve if they follow Dominance. But that 50–50 or so performance in two-player lab Disharmony and Partial Harmony games falls far short of overall HJV.

To get HJV for the class—to succeed in the Harmony equivalent of guessing the number of jelly beans in a jar or the weight of the ox—my business students had to do something very tricky. They had to balance out between the "greedy/needy" picking high numbers and the "generous/well-off" picking low ones so that the average pick of the class was just under 4. If the average nudges up to 4, everything collapses, with everyone getting only 1 point. But if the average is low—say 1, in a class of all self-sacrificers—the total number of points earned by the class also goes down, with few or no people getting 3 points.

The five Rutgers classes in which I've done the survey so far have all aced it. They spread themselves between high and low picks in such a way that the average in both classes was nearly 4, but just under it. Was it luck, a sign of the magic that can be achieved through the ethical wisdom of crowds, or something in-between? I'm very interested in whether these results are replicated in other classes, whether mine or other teachers', with nonbusiness students, with groups of strangers, and with money, either for oneself or for others, at stake. One straw in the wind: The one population I've tried the survey on that wasn't successful in getting a high payoff consisted of second and third graders in a religious education class, who picked high numbers, resulting in low payoffs for everyone.

With the wisdom of crowds exercise, like all the others, there is room for normative discussion, or debate. The results I've gotten so far, with the five classes of business students achieving very close to HJV by just barely avoiding disaster, impress me—but they also remind me of Icarus flying high until his fatal crash. With the heights we have reached as a species through our great skills in Harmonizing comes the danger of a fall; the "go broke at 4" game can be seen as a parable both of our brilliance, and of the risks it can engender.

Ethical Types: Who Leads?

Our final game turns from Harmony in general to its Sanguine/Happy, Melancholy/Norm-Following, Choleric/Competitive, and Phlegmatic/Pragmatic forms. The intuition I've tried to test in the "Who Leads" game is that the quality of cool, calm, calculation, central as it is to business ethics, is *not* a style that makes one likely to be a business leader, compared to the other three styles. The method I've pursued is to ask students—and also executives at a New Jersey bank, and professionals—to read descriptions of four character types: a happy type who thrives on challenge, a turn-the-other cheek type who forgives challengers, a competitive type who gets mad and punishes challengers, and a calm, rational type who calculates costs and benefits. I've then asked the respondents to assess which type is closest to their own style, and then to consider whether they would prefer to lead or follow if matched with a person of a given type, in a situation in which they are told the best outcome is from one player leading and the other following.

The reasons for my expectation that a more emotional character type will be more likely to lead than the cool, pragmatic type basically go like this: A Sanguine, challenge-loving type will be viewed as likely not to be disturbed by both players trying to lead, and is hence likely to lead. Knowing that, a Phlegmatic, cool, calculating player will follow. Similar reasoning applies to the forgiving player, who will likewise not be bothered by a leadership challenge, and will thus also be understood by the Phlegmatic player as likely to lead. A Choleric, angry player will be bothered by being challenged, but also will be deferred to some degree by the Phlegmatic player, I expected, based on a desire to avoid unpleasantness.

The differences I've found so far do indeed go in the predicted direction, in that people matched up with a Phlegmatic or "rational" partner are more likely to lead than people with other kinds of partners. The effect of people's own type compared to their partners' type is smaller, but has also gone in the predicted direction. One case—vivid for me, though only one person—involved the top executive I've dealt with so far, the head of the bank I presented the survey to. He pulled me aside after my presentation to tell me about his studying psychology as well as economics as an undergraduate, and that he recognized himself as a happy, challenge-enjoying leader first, and, then, after a bit more reflection, as a forgiving one second. The small number of executives and professionals in my sample who shared his "happy-forgiving" combination of styles all said they would lead, which was a statistically significant difference between them and their peers, most of whom put "rational" as either their #1 or #2 style.

Put broadly, what I've been interested in testing in Who Leads is a variant of Schelling's basic notion that a character type that allows you to make credible commitments can be a significant advantage in playing a game.[27] Whether leading as opposed to following is actually advantageous can be argued either way, based on assumptions one makes about moral emotions as well as other factors. I have set up my games with advantageous leadership so far, but plan to try the reverse framing in which leadership is a sacrifice, and also a "no numbers" framing, included in the appendices. My expectation is that the more emotional, less calculating character types will also be more likely to lead in these framings, perhaps to a greater extent than in my original one. To loop the discussion back to moral emotions and ethics: Business ethics, much as it can be understand as a role ethics that advances the Phlegmatic, pragmatic quadrant of ourselves—an argument that is central to the following, final chapter—is also a system of practice in which ethical types other than the Phlegmatic are well represented and successful.

* * *

On the day of our presentation to the anti-prostitution abolitionists, I had cautioned Sasha to stay away from risky language; I didn't want to offend anyone. When he mentioned how providers and sex workers used the Internet to post their calendars of availability and their reviews and ratings, the leader of the group, G _____, became visibly uneasy. I watched her and I knew he was about to get into hot water.

Then something unexpected occurred. At the end of his presentation, I was eager, anxious to do damage control, but a few participants approached him first. I listened in. Someone said to him: "Thank you for coming. If it weren't for your presentations, this would have been just another meeting of us talking about the same things over and over again." I realized in that moment that in trying to control the words, I had gone too far—I had limited the space for honest interaction and reflection. I was contributing to the maintenance of the polarized discourse. Maybe Sasha was right. Maybe I should lighten up.

—Nicole Bryan, Academy of Management paper and presentation (2014)

Harmony and Manipulation

One must be able to face and reflect upon the dark side of one's ethical construct, rather than simply trumpet its bright side. Just as other visions have their conundrums and their shadows, so too with Harmony. For example, as discussed in Chapter Three, a sense that the other player is failing to understand that the game is Harmony and is acting irrationally may well make one madder than a sense that the other player is acting egoistically. For another, as I discussed in Chapter Two, and as I return to here, Harmony games in some cases are not good overall, even if their element of Harmony is good; indeed, they can be embedded in serious wrongdoing and evil.

The issue I want to focus on here is a the connection between the creation of Harmony games and the manipulation of situations and people. As an intuitive Harmonist, one is always playing out scripts. As a reflective, System 2 Harmonist, one is always writing them. One may accept, and for that matter embrace, the Harmonist's play-acting and play-writing role—but one should acknowledge it, and be ready to accept that not everyone will find it his or her ethical cup of tea.

Consider the excerpt above, in which a colleague and coauthor of mine in critical business ethics describes how the presentation she and her husband made to a group of passionate antiprostitution activists went unexpectedly well—how, in the parlance here, a pleasant form of competitive Harmony prevailed, with clear differentiation between, but also respect, and perhaps affection, between, her husband, the outsider male, and the insider female activists. The man, the female activists, and likely my colleague, for all the self-deprecation of her account, wound up doing a good job at System 1 Harmony.

A good System 2 Harmonist will learn from experience, and experiment with new scripts going forward. As an advocate of Harmony, I am very much fine with that. In particular, I am very much in support of the idea of flipping the script as an idea for my colleague, her husband, and all of us to follow. I endorse the idea of treating one's life, and one's interactions with others, in terms of brief Harmony games of different kinds that one creates through the use of System 2 reason, as well as System 1 intuition. At the same time, I am fine with the idea that others with ethical sensibilities as refined, or perhaps more so, than mine may have a negative reaction to what they may regard as emotional manipulation, and may regard my favored approach as ethically questionable.

Manipulating Harmony Rather than Defying Harmony as a Way to Achieve Ethical Ends in Firms

As an example of manipulating Harmony scripts in a manner that is fairly extreme, but that seems to me warranted under the circumstances, I use the WorldCom accounting fraud case I mentioned at the beginning of the chapter.[28] The issue, phrased in terms of critical business ethics, is whether one can lead in reshaping troubling Harmony games, rather than simply acquiescing in them, or resorting to Disharmony. The scenario in a nutshell: At WorldCom, a sprawling telecom conglomerate that included the long distance carrier MCI, the word came down to accountant Betty Vinson from the company's CEO Bernie Ebbers, through CFO Scott Sullivan and a few layers of management, that she was to commit fraud (by capitalizing line costs that according to both common sense and generally accepted accounting practice are clearly expenses) to make the company's earnings look respectable rather than disastrous.

In WorldCom, as in many other canonical business ethics cases involving corporate cultures gone wrong, a natural way to think about the situation is that one faces a choice to be loyal to the norms of the firm as determined by its leaders, or to disobey those norms based on loyalty to another, better, more universal set of norms. One is like Milgram's subjects, faced with a choice of whether to obey the immoral command of the experimenter or to disobey it. Or, in the terms used here, one is faced with a group that expects you to play a bad version of a Harmony game. One can acquiesce in playing the bad Harmony game, or one can defy it by playing another game in which the players have different interests, such as a game in which you threaten to bring the company down by going public unless it reports its earnings honestly, or a game in which you mistrust the company and go it alone by going directly to the authorities.

I don't think the "fold or fight" way to think of the situation with a corporate culture that has gone off the rails is wrong at all. In my MBA classes, I reinforce that way of thinking by teaching WorldCom and the Milgram experiments in the same class. But I also believe that the Harmony perspective helps suggests an alternative to fighting or folding for Betty Vinson, her managers, and others faced with ethical dilemmas. Instead of defying a bad

Harmony game, or acquiescing in it, one may be able to lead by setting up a better Harmony game. One may be able to draw on people who can help, on one's social skills in bringing those people together with you and other people who like you are part of the off-the-rails group culture, and on one's ability to plan and to intuitively script a social situation.

At WorldCom, the company's own internal audit unit shut the fraud down when it learned about it. But it was too late for the company, and also for Sullivan (who claimed that CEO Bernie Ebbers had initiated and masterminded the fraud and received a five-year sentence), Ebbers (who went to trial and received a twenty-five-year sentence), Betty Vinson (who pled guilty and served a five-month sentence), and Buddy Yates (Vinson's manager, who received a year-and-a-day sentence).

The specific thought here is that given the will and the social skills, WorldCom managers and employees from Sullivan as CFO down the ladder to Vinson (as an accountant making the fraudulent entries) might have been able to manipulate their situations to create possibly duplicitous but pro-social Harmony games as an alternative to committing fraud, or to defiance. For example: At the top of the ladder, Sullivan could have cited a staff revolt, or another made-up factor, to say something to his CEO like, "I'm with you all the way, Bernie—screw the f-ing goody-goodies—but this just isn't going to fly. Here's what's happened since we talked last." Ebbers could challenge Sullivan's deceptive claim, of course—but he will be under strong Harmony pressure to accept Sullivan's credibility and to participate in Sullivan's new version of a Competitive Harmony game.

At the bottom of the ladder, Vinson could have pretended to be naïve by bringing up the fraud in an apparently off-the-cuff way in a meeting she was having with an internal WorldCom auditor for another purpose: "Oh, just one more thing . . . Buddy keeps explaining to me how this line cost business works." If her Melancholy Harmony game with Vinson and the internal auditor, and then others, in the company works, the auditor and others know that in fact she is savvy in dealing with the issue in the way she is, rather than through acquiescence or defiance, and respect her for that, and the information she passes on leads to the fraud being nipped in the bud.

Duplicitous or quasi-duplicitous Harmony games like those sketched out for Sullivan and Vinson are ethically imperfect, to be sure. The one sketched out for Vinson is troubling in relying not only on deception, but also on undesirable and dated negative stereotypes about women as naïve or stupid, and the one for Sullivan has him lying about, or at least being manipulative about, his own feelings, as well as about the facts. But both games, for all their real problems, are pro-social Harmony games that would have been very much better alternatives in their basic nature, and quite likely also in their consequences for the individuals and their company, than the corrupt Harmony games the WorldCom employees actually played to the bitter end.

I acknowledge and affirm that Harmony is not always the right game to play. Sometimes blowing up Harmony is the better course of action. Especially under the clear-cut WorldCom facts, I believe that most of us, at least in the United States, would admire a bold and defiant Betty Vinson who stands up to the company (by writing a letter to Ebbers, Sullivan, internal audit, and the firm's external auditors and counsel detailing an ongoing, multibillion dollar accounting fraud) more than we would admire the clever, manipulative, and socially skilled Vinson of my scenario. I would not, but I am also fine with being in a minority on that score.

Though I do not expect to reverse deep-seated ethical intuitions, I hope that this discussion has made the following proposition more of a live possibility for you as my readers: Employing the strong norms of Harmony that govern human social interactions in business, as in other settings, to create a new, better Harmony game is all too often overlooked as a possibility in situations in which we feel pressured to be part of a bad Harmony game.

A final reason for considering a Harmony strategy in business ethical dilemmas, and other ethical dilemmas, is worth noting: The moral reality of a given situation, obvious though it may be in WorldCom and some other situations, is very often—usually, I would suggest—anything but clear. For an ethically troubled employee who worries that she is being pressured into a bad Harmony game, but who also wonders whether the game is in fact bad—suppose Vinson in another set of facts is not in fact sure that the accounting treatment her bosses want is in fact wrong—her uncertainty about the state of the world provides an additional important reason for her trying to lead as a low-ranking employee by turning Harmony in her favor, rather than abandoning Harmony for defiance, mistrust, or resentful acquiescence.

From Micro to Macro

In this chapter, I have tried to relate the moral emotions, Harmony-centered version of game theory advanced in the first four chapters to the day-to-day practices of teaching, studying, and writing about business ethics, and to the classic business ethics issue of entanglement in an immoral corporate culture. The next task is to consider whether the game-theoretic approach taken here suggests a different definition of what business ethics is from the ones that we usually adopt. I believe it does; in the next and last chapter I make that case (Figure 5.1).

Critical Business Ethics • 139

	Does the Actor Intend the Consequences? Type of Actor			
Type of Consequences	Egoistic + Calculating (Knobe CEO)	Egoistic (child)	Altruistic (Green CEO)	Altruistic + Calculating (Benthamite CEO)
Positive Consequences	No	No	Yes	Yes
Negative Consequences	Yes	No	No	Yes

Experimental philosophopher Joshua Knobe found that his respondents believed that a profit-seeking CEO who caused harm to the environment intended those consequences, while not intending benefits to the environment caused by profit-seeking. The matrix includes the Knobe effect, as it is called, in the first column, followed by hypothesized effects in the latter columns that I have been testing with my business ethics students. The thought is that Knobe-derived experiments, along with experiments derived from the work of Thomas Schelling, Robert Frank, and others, can be of value in understanding the nature of moral intuitions and moral reasoning.

Figure 5.1 The Blame Game

Summary

This chapter explores connections between the Four Temperaments approach to game theory and teaching, research, and practice. The largest part of the chapter consists of a discussion of various ways in which a moral emotions approach to game theory can be incorporated into teaching with classroom games, discussions, and written work. I also consider possible implications for researchers, and for practicing managers. The focus is on business ethics, but the material is intended to be of use as well to teachers, students, and practitioners in other fields, such as management, supply chain management, economics, psychology, and law. The central claim is that the Four Temperaments approach lends itself to a critical approach, or school, that is receptive to, and committed to, both science and the humanities. In such a school, the litmus test for inclusion is not left–right politics, or deontological–utilitarian–virtue ethics, but methodological inclusiveness. Those who embrace both science and the humanities equally, even if their personal proficiency in one domain is much greater than in the other, should consider the proposed critical approach to business ethics, and by extension to other applied disciplines, a congenial home for their teaching, research, and practice.

Exercises

1. Write a team paper in which the members of the team express themselves in their own voices, and in which Sanguine Harmony, Choleric Harmony, Phlegmatic Harmony, and Melancholy Harmony are all present, along with Disharmony.
2. Write an individual or team paper in which there is a balance among the elements of science, arts, and practice.
3. Locate an academic or practitioner paper (or a section of a book) that, like the Cable, Gino, and Staats paper discussed in the text, supports the value of a humanistic management approach; relate that paper (or book section) to the perspective on Harmony and work discipline discussed in the text.
4. Write an ethical relations, moral emotions analysis of a business decision, using the text and the sample analysis in the appendices as starting points.
5. Try the following games on yourself, someone you know, or a class: (a) The Blame Game; (b) Ethical Focal Points; (c) Judgments of Unfairness; (d) The Ethical Wisdom of Crowds; (e) Who Leads?—see text and appendices.
6. Write a paper in which you describe a business or other social situation you were part of; describe how you could flip or alter the script to play Harmony instead of another game, or play a better version of Harmony.
7. Pick a business case; draw from the WorldCom discussion in the text to consider how an actor in that case might use manipulative tactics that would preserve Harmony, and that might achieve ethical ends better than defiance, exit, or acquiescence. Consider whether the actor's doing so is appropriate in your view or not.

CHAPTER 6

Why Business Ethics Matters

> Arjuna: I do not see how any good can come from killing my own kinsmen in this battle, nor can I, my dear Kṛishna, desire any subsequent victory, kingdom or happiness. Of what avail to us are a kingdom, happiness or even life itself when all those for whom we may desire them are now arrayed on this battlefield?
>
> Krishna: While speaking learned words, you are mourning for what is not worthy of grief. Neither he who thinks the living entity the slayer nor he who thinks it slain is in knowledge. There is neither a slayer nor a slain. One who has taken his birth is sure to die, and after death one is sure to take birth again. Therefore, in the unavoidable discharge of your duty, you should not lament.
>
> —The Bhagavad Gita

In our lives, there are times to make peace, to make love, to make amends, to make money, to make tracks, and to do many other things. There is also a time to make war. Making war in the form of making a systematic, reasoned argument that A is better than B, in a situation in which other reasonable people can, and should, argue that B is better than A, is not for everyone. But for those of us who, in at least part of our lives, are politicians, lawyers, normative philosophers, or other types of advocate, it is our duty.

As a member of the word-wielding warrior class, the battle that one fights may be against the literal wars of human tribes with arrows and chariots, and for the moral equivalent of war, for practical forms of Choleric Harmony that are as brilliant, as sad, and as angry as what we have always known as a species, but in which no one kills anyone. That is, I believe, the war that Joshua Greene calls for us to enlist in at the end of his recent book *Moral Tribes*, when he turns from moral psychology to normative philosophy: If we can all embrace liberal, universalist utilitarianism, we can conquer our bloody tribal history. Or, alternatively, one may fight a battle for what one sees as better balanced forms of Choleric Harmony, ones in which the self-righteous spirit and the coercive, compliance-compelling spirit in argumentative advocacy are tempered to a greater degree than they typically are by the spirits of calm and happiness. That is, I believe, the battle that Jonathan Haidt is fighting in *The Righteous Mind*. If we are truly to live up to the promise of universalism, we must open our hearts and minds to traditionalist, conservative particularism.[1]

Greene's overt normative argument and Haidt's implicit normative argument about moral tribes are not related in terms of business, or of business ethics, much as they have implications for both. In this chapter, I draw on the Harmony perspective advanced in the earlier chapters to make a normative argument about the place and the future of business ethics that has human moral tribalism as its background. I argue for the following three core claims: First, an understanding of human history and the human present in terms of Phlegmatic business ethics on one hand, and Melancholy, Choleric, and Sanguine forms of ethics on the other hand, or hands, is better in some significant ways than major alternative understandings, such as those offered by Hegel, Marx, Nietzsche, and Hegel's modern interpreter Francis Fukuyama. Second, as business ethicists we should work to make our particular version of ethics the best that it can be, while at the same time upholding the equal dignity of the other major forms of ethics, and avoiding claims about the moral superiority of our form, or of any other form. Third, we should aspire to a future in which business ethics is more of a universal possession, and less the intellectual property of managers and other elite groups, and in which Sanguine ethics is more on a par with, or somewhat above, its Phlegmatic, Melancholy, and Choleric peers.

Two Perspectives on Why Business Ethics Matters

In one view, which I believe to be the orthodox one, business ethics matters both because business, like all human activities, ought be conducted ethically, and because the intensity of pecuniary motivation in business brings with it certain distinctive ethical issues. In the different view advanced here, business ethics matters both because it is the ascendant ethical system of our time, and because its relations with other ethical systems that were formerly ascendant are uneasy, and are capable of being improved.

The standard perspective on why business ethics matters goes along with a sense of business ethics as weak, or at any rate as weaker than it ought to be. The alternative perspective advanced here goes along with a sense of it as strong both in itself, and relative to its temperamental peers. The standard sense of why business ethics matters corresponds to mainstream business ethics; the alternative sense of why it matters corresponds to a potential critical business ethics movement. Such a movement, in addition to, or apart from, engaging in the projects that were discussed in the last chapter, could engage in normative scholarship and practically oriented writing that starts from an assumption that as a business ethicist, one stands at the center of power in the contemporary world, rather than in a peripheral, plaintive position.

The conventional understanding of business ethics as an underdog corresponds well to the practical situation of business ethics as a small academic field. The annual meetings of the Academy of Management draw thousands of academics; the annual meetings of the Society for Business Ethics, held in the same city at the same time, draw a very much smaller number. To accord with reality, the claim advanced in this chapter about the ascendance of business ethics needs to be understood not as a claim that now, or anytime in the foreseeable future, business ethicists will outnumber management professors on business school faculties, or that people in the workforce with job descriptions as ethicists will outnumber people with job descriptions as managers. Rather, the claim here needs to be understood as a claim that the system of fused beliefs and practices that both management and business ethics professors, as well as the rest of our business school colleagues, uphold and seek to improve, is, more than anything else, what is distinctive about the world we live in now. We may be, and I believe we should be, open to Sanguine as well as other feelings about our social reality—but it is good, I believe, to see the power of our discipline as one truth about our situation, even if at the same time we also believe in, as we logically may, the traditional story of business ethics as in need of strengthening, and of ourselves as weak. We in business ethics may be tiny, but the Phlegmatic, pragmatic worldview that it is our role to expound and perfect is the regnant system of our time.

In the Four Temperaments perspective that has been developed in this book, human beings have been very good at solving social games from our earliest days as a new, struggling species in Africa to the present. Then and now, our shifting, mutable moral emotions help us solve Disharmony and other difficult games. Then and now, the Sanguine, the Phlegmatic, the Choleric, and the Melancholy all have important roles to play. At the same time, it is reasonable to believe that there have been significant changes over time in how we solve social games. It is plausible that the way we solved games over our many thousands of years wandering the savanna in small bands was in some significant respects different from the ways we solved them in the cities and villages of the first agrarian civilizations in the river valleys of Asia and Africa. It is further plausible that the way we solve them now, in the towering cities and sprawling suburbs we have built around the world over the past few centuries, is in turn significantly different from the ways our agrarian forebears solved them. As a broad hypothesis, or conjecture, for further exploration I would suggest the following: Human history can be understood as the history of how we solve social games. As a narrower, more provocative, one, I would suggest the following thought, to be developed in the next section: Human history can be understood, in significant part, as the history of change in solving social games in three main material orders, one based centrally on hunting and gathering, another based centrally on agriculture, and the third based centrally on business.

* * *

> The truth of the independent consciousness is accordingly the consciousness of the bondsman. [J]ust as lordship showed its essential nature to be the reverse of what it wants to be, so, too, bondage will, when completed, pass into the opposite of what it immediately is: being a consciousness repressed within itself, it will enter into itself, and change round into real and true independence.
>
> —Georg W. F. Hegel, "Lordship," §193, *The Phenomenology of Spirit* (1807)

> The slaves' revolt in morals begins with this, that ressentiment itself becomes creative and gives birth to values: the ressentiment of those who are denied the real reaction, that of the deed, and who compensate with an imaginary revenge. [T]he reverse is true of the noble way of evaluating: it acts and grows spontaneously, it seeks out its opposite only to say Yes to itself still more gratefully, still more jubilantly.
>
> —Friedrich Nietzsche, *The Genealogy of Morals* (1887)

> The hand-mill gives you society with the feudal lord; the steam-mill society with the industrial capitalist.
>
> —Karl Marx, *The Poverty of Philosophy* (1847)

> Do I obey economic laws if I extract money by offering my body for sale, . . . —Then the political economist replies to me: You do not transgress my laws; but see what Cousin Ethics and Cousin Religion have to say about it. My *political economic* ethics and religion have nothing to reproach you with, but— But whom am I now to believe, political economy or ethics?—The ethics of political economy is *acquisition*, work, thrift, sobriety—but political economy promises to satisfy my needs. . . . It stems from the very nature of estrangement that each sphere applies to me a different and opposite yardstick—ethics one and political economy another; for each is a specific estrangement of man and focuses attention on a particular field of estranged essential activity, and each stands in an estranged relation to the other.
>
> —Karl Marx, *Economic and Philosophical Manuscripts of 1844*

Idealist, Materialist, and Fusionist Accounts of History

If we are to develop an alternative way of seeing in which business ethics is a world-spanning field, we need to be able to situate it in a philosophical history. Toward that end, I suggest in this section that the wide-angle visions of historical stages in Hegel, Nietzsche, Marx, and Hegel's modern exegete Francis Fukuyama can be usefully drawn on as well as criticized for their limitations. To anticipate the conclusion of the argument in this section: Rather than seeing the central ethical truth of the present age as the ascent of freedom and democracy (Hegel and Fukuyama), the ascent of slave morality (Nietzsche), or the ascent of a material order of phenomenal productivity that entails a new ethics of communist sharing (Marx), one may reasonably see the central moral reality of our time as the ascent of a Phlegmatic, pragmatic, business ethics mode of solving social games.

First, Hegel, whose early nineteenth-century theory of history as a progression toward freedom is the one that, more than any other, reflects the interests and the value commitments of contemporary educated, affluent Westerners, viewed as a class. His Berlin lectures on history, phenomenally popular as they were in his own time, do not read especially well now.[2] There is not enough philosophy to make the message sing in the way his master–slave dialectic in *Phenomenology of Spirit* does, and the emphasis on German—or more broadly, northern European—people as the planet's avatars of freedom, is redolent with moral tribalism by modern standards. But Hegel's vision of a progressive transition over the millennia from an initial stage of Freedom for One—which he sometimes calls Oriental Despotism, and which may be identified historically with the emperor system in China—to a stage of Freedom for Some—which he identifies with Aristotle's and Plato's Greece, with its combination of democracy and slavery—to a culminating and final stage of Freedom for All—which he identifies with Christianity and northern European peoples, but also views in broader, universal terms, as the end of history, the destination at which all peoples will arrive—remains, I would suggest, profoundly important today, even as—and partly because—it provokes unease and controversy.

Although Hegel is a dim figure nowadays, the Hegelian interpretation of history as progressively directed is very much with us. Fukuyama's *The End of History*, with its vivid closing image of the wagon train of history leading up to liberal democracy, is self-consciously Hegelian philosophical history that is both powerful and provoking in its claim of universality. Greene's *Moral Tribes*, with its depiction of moral progress over time and its brief for further progress through the adoption of liberal, universalist utilitarianism, is grounded differently, but shares with Fukuyama a progressive, neo-Hegelian vision of history; Steven Pinker's detailed chronicling of the diminution of violence over time in *Better Angels*, though not self-consciously philosophical in the way Fukuyama and Greene are, has a similar spirit.

Just as Hegel had his nineteenth-century critics—we will turn to the two most famous ones shortly—neo-Hegelian universalism has its contemporary critics, who assert the contemporary and future viability, if not necessarily the normative desirability, of human moral tribalism. In addition to Haidt's criticisms of liberal moral triumphalism in *The Righteous Mind*, a notable example of a reaction to modern neo-Hegelianism is Samuel Huntington's *The Clash of Civilizations and New World Order*, with its pessimistic vision of unending, intractable cultural conflict. Though my own sympathies in the factual "Has the world gotten better and is it likely to do so in the future?" debate lie with optimists like Fukuyama, Pinker, and Greene, in the ethical "Are we better now/in the West/on the liberal side?" debate, I believe the better case lies with critics of progressive triumphalism like Haidt, and, to the extent he is interpreted as an anti-Hegelian moral critic as well as a Hobbesian realist, Huntington. An Hegelian, or neo-Hegelian, vision of those of us who are citizens of stable liberal democracies, or who are liberal universalists, as having arrived at the perpetual summit of world history, inspiring though it is in its optimistic sense of tribalism transcended, is deeply troubling in its own tribalism, in the lines it draws between the enlightened and the left behind.

Let us turn now from the Hegelian sunny side to consider two dark portraits of modern ethics. In similarly Choleric, but otherwise very different, fashions, Hegel's great nineteenth-century critics Nietzsche and Marx flipped his story of history reaching its culmination in universal freedom into pessimistic stories of resentment and exploitation as the prevailing spirits of modernity. In Nietzsche, Hegel's account of the present as the era of universal freedom and the classical era as the era of Freedom for Some, with its accompanying moral condescension toward the past and toward traditional cultures, is replaced by a bold, splenetic redescription of human moral history. The dependent Lord of Hegel, in thrall to his bondsman, becomes Nietzsche's Sanguine noble soul that calls out "Yes!" to all, including its detractors and foes. Nietzsche performs a parallel flip on the present. We moderns are not Hegel's free souls who have transcended the master–slave dialectic, but are rather ashamed, slavish souls who over the past few millennia have channeled our unhappy consciousness into an ever-expanding roster of resentment-laden moralities, beginning with Judaism, Catholicism, and Protestantism, and yielding in modernity to a new, ever-multiplying list of rationalized systems of resentment that includes Kantianism, utilitarianism, liberalism, socialism, anti-Semitism, and German nationalism.

Given the constant, surging, Choleric bile in Nietzsche's highly self-aware portrait of modernity, it is hard to miss the applicability of his criticism of resentment to his own thought and feeling. He is, one fears, the prisoner of the very emotion he sees as the demon of the modern world—or he is, one hopes, an avatar of affirmation for all of us, and for the messy, ashamed, hopeful, angry moralities of modernity. In either case, he embodies his own system in a way that is simultaneously appealing and off-putting. In him, philosophy has started telling the embodied truth about the connection between our emotional, intuitive System 1 selves and our reasoning System 2 selves, one feels—but one also feels that in his case the self corresponding to the system is too Choleric, too disturbing, too far from a healthy balance among the temperaments, too lacking in steadiness and calm. One may agree with him—I do—about the predominance of a deeply appealing Sanguine element in the philosophy of a self-confident ancient like Aristotle that is missing, or much weaker, in the deadpan, sad, pessimistic, anxious, angry philosophies of the great moderns—but one is also left feeling that his portrait of slave morality, provocative though it is as a redescription of the ethics of the last two millennia or so, misses what is most distinctive about the past few hundred years of human history. To do that, we need, I believe, to paint a moral portrait of ourselves that has more to say about business, about life on the shop floor and in the office, than either Hegel or Nietzsche, with their idealist, armchair, traditionally philosophical stance, has to say. The history of our ethics is connected, one intuits, to the way we live our daily lives. To be plausible, an account of our moral situation needs to address the major material changes in our condition over the past few centuries. Our next great critic of Hegel, Marx, makes that a central article of his system; to his portrait of the history of human ethics we turn next.

Instead of starting with classical civilization, as both Hegel's and Nietzsche's portraits of our moral history did, Marx and his collaborator Engels emphasized as well the long hunter-gatherer period that preceded civilization. In that period, they saw an original, primitive communism of egalitarian sharing, along with much struggle and strife, all associated with a mode of material production in which armed male adults cooperated in hunting and sharing game that could not all be eaten by one person, or one family.[3] With the great change beginning ten thousand years or so ago to a new material order of agricultural production, and the need it engendered for long-term storage, command and control, and large-scale social institutions, came a new kind of ethical order, one in which egalitarianism gave way to obedience and hierarchy, with religious and moral codes promulgated by, and for, a relatively small ruling class of kings, nobles, priests, scribes, landowners, and major merchants, and applied upon the minds and bodies of large subordinated classes of agricultural workers, urban laborers, small-scale craftsmen, petty entrepreneurs, and, often, slaves.[4] Finally, with the modern change to a prodigiously effective machine-aided mode of production that has dwarfed earlier systems in its ability to create abundance for all has come a new ethics of communist sharing.[5] One may struggle for the success of that ideal—as Marx and his followers did in socialist and communist parties—but one should also recognize at the same time that communist sharing is the universal logic of the mode of production itself, not the sectarian ideology of a proletariat that is exploited by the last ruling class, the capitalists, and by their ideology of free trade. History and its unhappy class struggles will end with abundance and its accompanying ethics of sharing. With communism, we will be free, at last, from the chains of scarcity-imposed roles and duties for the first time in our existence as a species, free as we wish to hunt, fish, rear cattle, or criticize, without ever having to assume a job, or a single occupational role.

Hard as it is to abstract away from the practical program of state control of the economy in which Marx embodied his story of history, and which followers of his, such as Lenin, Stalin, and Mao, implemented in important parts of the world in the twentieth century, if we are to understand Marx's account of modernity, especially if we are to understand it in relation to the business ethics account proposed here, we need to be able to do so. Marx's story of history ending, of paradise around the corner, or at least in prospect, could have been—and can be now—accompanied by laissez-faire ideology, as Herbert Spencer's nineteenth-century story was,[6] or by an above-the-fray, twenty-first-century techno-utopian detachment from politics. What is most distinctive, and relevant for our purposes, about Marx, in comparison to his fellow big-thinking German historical philosophers of the nineteenth century, is not his credal attachment to the abolition of private property, but his preoccupation with the material. In his portrait of modernity, business is at the center of the stage. And in that, if not in the imbalanced Choler of his particular portrait, he is deeply right. If one imagines one single, great, room-sized, neo-Breughelian painting of life in the cities and towns of northwestern Europe in the mid-nineteenth century—one can think of it displayed in a circular room like the Orangerie in Paris, or the similar room in Gettysburg—that portrait would include some soldiers, legislators, demonstrators, and a legislative hall or two; some worshippers, prelates, and a church or two; some students and teachers, and a school or two; and some traditional scenes of peasants roistering and reaping. But at the center of that painting of the new order of the Western world, and in the background and at the edges of it as well, would be the urban laborers, the millhands, the clerks, the shopkeepers, the merchants, the owners, and their families, all arrayed in their many streets, factories, shops, and offices, and in their many shabby, middling, and grand homes, some of them in toil, some of them at ease, many of them in between, some of them alone, some of them engaged with one another. The painting would have business and industry at its heart, and everywhere.

We are now ready to identify what is distinctive about a potential business ethics account of history and of modernity, and how it is both indebted to, and different from, Marx's account. Like Marx, a business ethics account places business at the center of the history of our time. But it semi-flips Marx's one-sided materialist flipping of Hegel's idealism, and arrives at a view of business as a fusion of the technological/inhuman/material with the value-laden/human/ideal, with neither side ruling over the other. Instead of treating the ethics of "*acquisition*, work, thrift, sobriety" that Marx identified with political economy as superfluous persiflage, or as an estrangement of man from his true being, or as a rationalization of class interest, the account of modernity suggested here takes political economy ethics—that is, business ethics—to be both the ascendant ethics of our time, and a central contributor to modern order of high material productivity.

Are we now back to Hegel, with "business ethics" substituted for "freedom" as the great spirit of the age to which we should bend our knee? Not at all. Phlegmatic, pragmatic, practical business ethics, important and valuable as it is, is a role ethics, just like the Choleric, egalitarian, warrior ethics ascendant in our very long hunter-gatherer era, under which, in Boehm's and other recent accounts, armed men shared their game and resisted bullies in their ranks,[7] and the ashamed, Melancholy, hierarchical, priestly ethics ascendant in the agricultural era. Business ethics is not the end of history at all. It is one stage, our stage and important withal, but not the final destination of history.

Beyond the modes we know now, there will be other modes of production in the future—perhaps the extension of human life to hundreds, or thousands, of years, perhaps mass production under the control of individuals,[8] perhaps the rise of machines that will write works far more interesting than the present one—and new future ethics that will help engender those modes of production, and will in turn be fostered by them. At the same time, beyond these and other contingencies, there is, we may reasonably believe, a logic to history under the account proposed here.

One may see history, including the history of our time, as marked, and also marred, by imbalance toward one or another of the Four Temperaments. Such imbalance is unhealthy for the flourishing of individual people and things, including, we may surmise, corporations; it has also, we may reasonably suppose, been unhealthy for the flourishing of humanity. Given that, one may aspire, through politics or otherwise, for a better balance among the four great emotion-based ethics in the future history of humanity, and in particular for the rise of the Sanguine to a stronger place in our feelings and in our reason than it now occupies. Arjuna had his time, Krishna had his, and now a different, practical, smithian spirit—India's Tvastr, Rome's Vulcan, Scotland's Adam—rules us. All very well—but we may wish, if we like, for a new goddess to walk among us, and to lead us.

Against Universalist Morality and for Temperamental Ethics

The idea of right and wrong is inscribed within us all, we may stipulate. A universal moral language is, among other things, a Schelling focal point that helps us to coordinate by creating Harmony games with one another, rather than being under the sole dominion of the often difficult games of Nature. But that does not mean that the language of a universal morality of right and wrong is, at a given historical moment and in a given setting, the right language for us to think in, speak, and write. At some such moments and settings, including, I would say, the present one, the time is right to speak out against universalist morality, and in favor of multiple ethics.

In universalist morality—whether that universal morality is Hegel's freedom, Greene's liberal utilitarianism, or something else—lies moral tribalism. In the ethics of different temperaments that recognize themselves as limited and tribal, as in constant deep need of balance, and of companion ethics—of Choleric warrior ethics, of Melancholy priestly ethics, of Phlegmatic business ethics, and of the Sanguine spirit that was more in evidence among classical philosophers than it is among us, and that may, one hopes, rise again in a future ethics—lies ethical universality. Without exalting our current era of ascendant business ethics as an ethical summit, or as an end of history, we may reasonably see it a time in which the logic of multiple ethics will become more accessible to us than it was in the preceding era of ascendant Melancholy, priestly ethics.

The spirit of compliance that is central to Melancholy ethics carries within it the logic of universal morality: "This is a command for all! Obey!" That spirit remains very much part of us now, as it should be, but it no longer has the compelling force over the cooler, Pragmatic spirit that is also a crucial part of our ethical selves that it once did. In our practice of Schelling's egonomics,[9] we can tell the ashamed, guilty side of ourselves: "We appreciate you. In many cases, you are exactly the stick we need to beat ourselves. We cannot do without you. Thank you very much! But on the specific issue of whether we should feel ashamed and guilty if we do not like the idea of a universal morality, or, rather, your all too particular, all too partial version of it, sorry! You've been outvoted by our Phlegmatic, Choleric, and Sanguine members. It is fine, in fact it is, we think, a splendid thing, to live a life in which we try to make our ethical temperaments and our ethical roles as good, as beautiful, and as true as they may be, without worrying for a minute—for even a split second!—about whether there is an overarching or underlying universal morality. And yes—we understand that the vote will come out the other way for some other people than it does for us, with universal morality supported by their Choleric as well as their Melancholy sides, and with only their Sanguine or Phlegmatic sides, or parts of them, dissenting. That vote may well be right for them, just as we hope our vote is right for us; the particulars of background, circumstances, opportunities, and challenges will be determinative. So let's figure out how to work together, shall we?"

For Political Ethics and Penitential Ethics as Temperamental Ethics, and Against Them as Master Ethics

The negative response to claims of universalist moralities that has been suggested here is hardly a novel one in its anti-foundationalist, critical drift. It bears a strong family resemblance to Nietzsche, Marx, and their milder American successors in critique, such as William James, John Dewey, Richard Rorty, Stanley Fish, the 1920s–1930s legal realist movement, and the 1970s–1980s critical legal studies movement led by Duncan Kennedy and Roberto Unger.[10] As one who graduated from the law school that was the center of critical legal studies, at the height of that movement, I am happy to avow its influence on me, which has been very considerable, as well as that of Rorty, whose *Philosophy and the Mirror of Nature* was a book I was very devoted to in the 1980s. What I would assert, though, based in part on my own experience, is that an anti-foundationalist mix of political Choler and Melancholy self-criticism is no more a master key to the universe than business ethics is. In particular, Choleric politics of whatever ideological stripe is not the answer; it is only one, limited ethics.

As with Choler, so too with Melancholy. Religious, legal, and ethical efforts to join us as one in a spirit of compliant humility are one extremely important part of ethics. But to confuse, say, obedience to the moral law, or to God's will, with the whole of ethics is a very serious mistake, just as confusing, say, libertarianism, communitarianism, liberalism, radical centrism, or another Choleric political ideology with the whole of ethics is.

To put the foregoing critical points in a more Sanguine fashion: Not the least value of the humoral perspective on ethics is that it gives us a good alternative to a Disharmony view of human nature as dichotomized between a morally worthy but, absent the big sticks of internal and external punishment, ineffective, altruistic sub-self, and a morally neutral, or unworthy, but effective, egoistic sub-self. In the humoral perspective on ethics, personal and social pleasure are conjoined, not opposed. Self-related and altruistic feeling coexist in the Sanguine quadrant of the self, and we are made so that the two types of feeling flourish and ebb together, rather than in opposition to one another. There is an important role for Choleric and Melancholy feelings in dealing with Disharmony and Partial Disharmony, and for the reasoned System 2 development of those feelings in political ethics, and in the penitential forms of ethics devoted to compliance with social norms—but one should not confuse these feelings, and the ethical systems that promote them, with the entire breadth of ethics.

Business Ethics and Traditional Ethics Contrasted

The view of business ethics advanced here bears a considerable resemblance, and a considerable debt, to Max Weber's portrayal of modernity as characterized by the idea of work as a calling. In what follows, I provide examples to illustrate the basic idea that business ethics can usefully be understood as an ascendant Phlegmatic, pragmatic, practical approach to solving social games that elevates busyness, industry, management, leadership, and followership as shared values and as material practices. The method will harken back to the earlier stories I gave of social games. In what follows, I present brief vignettes of how business ethics approaches to solving the major kinds of games—Harmony, Imperfect Harmony, Partial Disharmony, and Disharmony—differ from traditional ethics approaches.

Implicit in the vignettes is an argument about each of the games. For Harmony, the basic idea is that business ethics aligns us in what amounts to an adult version of children's parallel play: We Harmonize with one another by working alongside one another. For the equal payoffs version of Imperfect Harmony, my claim is that business ethics has a dual effect: It aligns us both in the direction of mechanizing and routinizing our behavior through adherence to rules, such as traffic laws, and in the direction of trust as a default assumption.[11] For the unequal payoffs version of Imperfect Harmony, business ethics works to align people toward shared management, leadership, and followership, and away from dependence on hierarchical roles. Finally, for Partial Disharmony and Disharmony, business ethics encourages us to treat the world as better described by the cooler, consensual, forgiving logic of Harmony and Imperfect Harmony than by either of the two difficult, dramatic, conflictual games. It pushes us to play Highest Joint Value ourselves, and to understand the player who fails to do so as ignorant, and in need of teaching, rather than as a bully, or a slacker, who needs Choleric punishment.

Phlegmatic Business Ethics Harmony Compared to Other Forms of Harmony

My wife and I recently visited the Crown Heights and East Williamsburg neighborhoods in Brooklyn, where she lived as a child. We then walked over to the gentrified neighborhood of Williamsburg proper, where we stopped in for coffee at Lovin Cup, which was filled with highly attractive young people. Recorded music played, but the atmosphere was as hushed as a library reading room. People on laptops, tablets, and phones manipulated their devices, without the quiet, or less quiet, voices and the squeals of the very young that are part of the café experience in well-off New Jersey suburbs like the one we live in. We were, it occurred to me, at Ground Zero for business ethics Harmony, viewed as a way of living and socializing, not simply as a mode of paid labor. The cafés of South Orange, Montclair, and Westfield that we were used to were filled with people quietly using laptops and other devices, but also with Sanguine, Melancholy, and, occasionally, Choleric activity. Business ethics Harmony was a central element there, perhaps the central one, but without the intensity and relentless focus of the cloistral Williamsburg café. The city bakeries of Newark and Plainfield that I enjoyed going to in my New Jersey journeys were a different kind of experience. I was the only person on a laptop there, and the patrons often lifted their voices louder in happiness, and sometimes in agitation, ire, and sadness. In the relations of the patrons and employees with other people, and with the world of objects, there was Harmony in all of the cafés, in vanguard Williamsburg, in middle-class New Jersey, and in working-class New Jersey alike. But it was Harmony of different kinds, rather than one.

Campaigning: Business Ethics Equal Imperfect Harmony

It is a rainy Saturday in early November, and I am walking my election district to hand out campaign literature to hundreds of homes on behalf of a school board candidate. The literature is tailored to appeal to people who have voted, and I am dropping it off only at homes in which one or more people voted last November, ignoring the others, including some homes of people I know well. I am not interested in talking to anyone, just in dropping off the literature. It is satisfying to turn myself into a machine, satisfying to tote people's water-soaked *New York Times* or *Wall Street Journal* to their front porches as an easily accomplished favor, satisfying to nod and say Hi and nothing more to the occasional people I see on the streets, satisfying to trust that I can do what I am doing safely, and with my role understood and accepted, satisfying to imagine other workers for my campaign and the other side's campaign doing their own versions of our shared ritual. The algorithmic, business ethics form of Harmony that I am experiencing with my neighbors and my community is more pleasant, it occurs to me, because last week I was doing something very different. I canvassed my own street and a few nearby streets door-to-door, ringing every bell and, depending on the interaction, asking people if they would put up a lawn sign for my candidate. In those up-and-down, mutable interactions there was Sanguine Harmony in a click of connection with a new neighbor who is a teacher in another district and was very happy to put up a sign, Choleric Harmony as I chat with a strong backer, Disharmony as a strong supporter of the other side and I fail to either deviate from our respective loyalties, or to advocate them with passion, and Melancholy Harmony in strangers' demurrals, accompanied by reasons like a need to study the literature first. All very well—but the absence of personal connection in what I am doing now on this rainy day, with its pure relation to things, to rules, and to a process, rather than to people and to their fluctuating feelings—is very nice, too. Phlegmatic Harmony is very appealing, in its place.

Shared Leadership and Shared Managerialism in Business Ethics Unequal Imperfect Harmony

The election has gone our way, and I'm imagining a conversation I'm not looking forward to having. I'm talking to the incumbent board president, explaining to her that I have five votes on our nine-member board to replace her as president. She is not happy. I give her a passage from John and Miriam Carver, management consultants we've been following over a number of years, on how board meetings should be lively engagements on policy rather than administrative presentations, and explain my hope for changing the way we do business as a school district away from shared managerialism to shared leadership. Now, the breakthrough moment in my imagined dialogue with her: As I soft-pedal my line in an effort to appeal to her, I—the real me—realizes that the social niceties about how her approach is just as good as mine embody a truth as great or greater than my own shared leadership line does. In my biased moral gut is a Choleric feeling, and an accompanying System 2 ideology, that the "everyone a leader and a follower," "break down the deadening routines!" approach to doing things on the board that I want to implement is far better than her approach. "The times are on my side! My approach is in keeping with an emerging anti-hierarchical logic of business ethics and business practice!" Perhaps . . . but, it occurs to me, her intuitive Melancholy managerialism, with everyone a worker bee, is at least as central a part of the logic and the daily practice of the pragmatic culture of business ethics as my possibly Sanguine, possibly all too Choleric hopes for self-aware, shared leadership and followership are. In that regard, it's interesting, it occurs to me, that there are many more nonprofit and for-profit boards that embody shared managerialism than the shared leadership and followership I hope to turn us toward. Maybe that is not foreordained; maybe there are inefficiencies in the ethical marketplace; maybe there are plenty of twenty-dollar bills on the ground to be picked up by creative leader-followers. I hope so. Come January and the beginning of my board presidency, we shall see, as my department chair at the business school likes to say.

The Subordination of Disharmony and Partial Disharmony in Business Ethics

Speaking of my department chair: As it happens, she has just been selected as the new Dean of my business school. As Vice Chair of her department, Supply Chain Management and Marketing Sciences, since shortly after its founding, I have worked very closely with her. If I can abstract from my happiness on her behalf, there is, I believe, a cool, practical, business ethics story to tell about her ascent. She, a Chinese-American engineer with an accent who came to America as an adult, has a skill in treating the situations she faces in terms of Harmony, rather than of Disharmony. Where I tend with my often Choleric disposition to say, irritably, "We should do X, right?!" she is more inclined to an equanimity, and an openness to the idea that multiple paths to a good outcome are possible. Partly through working with and observing her, I am intellectually committed to the position that the way forward for good leader-followers lies in creating Harmony, not in preconceived ideas about what the payoffs are in the game one is playing. But have I learned it in my bones? Now that she has moved up, can I embody it the way she has in new leadership roles in my job, and on the school board? We shall see!

* * *

Being Phlegmatic about the Phlegmatic

Business ethics is wonderful, I believe, in its alignment of human beings and things to achieve high individual, organizational, and social productivity. The young people silently working and playing on their devices in the Williamsburg café; my walking on in the rain to drop off the campaign flyers; a board and an executive united in shared leadership and followership, or for that matter in shared managerialism; people impelled by a shared sense of calling, and working together as though they are in Harmony, rather than partial or full Disharmony—these are all things of beauty. But there is beauty, and truth and goodness, in the Melancholy, Choleric, and Sanguine ethics as well.

Moreover, the other ethics, I would suggest, stand on a plane of moral equality with business ethics, with Sanguine ethics a shade above the three others.

In our framework, the game of games is Harmony. In the moments in different cafés in different places in which Melancholy Harmony, Choleric Harmony, and especially Sanguine Harmony prevail, there is, I would suggest, fully as much beauty, as much goodness, as much truth as there is in the business ethics moment at the Lovin Cup. The prodigies of productivity that business ethics engenders are highly impressive, to be sure. But productivity, seen through the calm eyes of the Phlegmatic perspective itself, is not something to bend the knee in worship to, any more than it is something to shake one's fist in Choler against. Nor is the internal psychic drive that is one very large part of that productivity. It is a wondrous thing that we in the business ethics era hold the gun to our own heads. But that discipline of our time, viewed calmly, is neither an ethical summit, nor an ethical nadir, when compared to the other great moral tribal forms of discipline, those of the warrior and the priest, that have reigned in earlier eras of human history.

One may account for another prominent feature of business ethics, its "no drama" preference for looking at human interactions in calm Harmony and Imperfect Harmony terms, rather than dramatic Disharmony and Partial Disharmony terms, in a similarly dispassionate fashion. The calm elevation of Harmony and Imperfect Harmony in business ethics works, one may plausibly maintain, in substantial part because law and social norms in certain cultures and subcultures make that position more advantageous to adopt than a traditional attachment to the dramas of Disharmony. It is worthy to see one's everyday world in Harmony terms, one may reasonably think, but if it is also what gets us ahead, it is not itself ethically elevated compared to the contrary, less advantageous Disharmony perspective.

Finally, how about the shared managerialism, or shared leadership and followership, of our era of ascendant business ethics? One may indeed put social equality above social inequality. I do, and I happen to believe—though it is not part of this book to argue the point—that there is a good Sanguine, Melancholy, and Phlegmatic case for doing so, as well as a Choleric one. But here we arrive at what seems to me a central moral tension within our ascendant order of business ethics, one that I believe a calm, Phlegmatic observer would conclude that it has thus far resolved no more successfully, though also no less successfully, than earlier ethical orders resolved their fundamental moral tensions.

In the anti-bullying, egalitarian, Choleric, warrior ethics of our very long hunter-gatherer era there was a fundamental tension—or contradiction, if you like—between the anti-oppression ethos that led armed men to band together to impose rough material and social equality among themselves on one hand, and the endemic violence directed at each other and at women that was (even for those of us who are comparative optimists about the human past) a central part of their way of life. Similarly, in the priestly, Melancholy ethics that rose as civilization rose there was a fundamental tension, or contradiction, between universalist claims of a divinely ordained moral order on one hand, and the reality on the other hand of particular material and moral orders ordained by real human beings, who stood at the top of those orders.

Now, in our era, there is a fundamental tension, or contradiction, between a business ethics commitment to social equality in the form of universal managerialism, universal leadership, and universal followership as effective means of solving games, and the reality that many, perhaps most, of us, do not feel ourselves, whether in our jobs, or in other spheres of our lives, whether as individuals, or as members of certain underdog groups, to be empowered managers, leaders, or followers in an engaging, energizing, inspiring way. Life is great for some of us—for Freud, for example, with his Sanguine love of work, and for certain others of us—but definitely not for all of us.

Choleric disagreement as to who is at fault for this state of affairs—"Managers!" "Workers!" "The left!" "The right!" "All of us!" "Nobody!"—we just need to accept necessary, ordinary human unhappiness!"—is, of course, possible. But there is another, less Choleric way to think about where we stand now in the era of business ethics, and about where we may be heading in the future. We may reasonably believe that Phlegmatic business ethics by its own cool lights is not morally better than Choleric warrior ethics and Melancholy priestly ethics, much as we may at the same time deeply appreciate the economic and cultural productivity it has unleashed. But we may also reasonably hope that its success in enhancing our productivity opens the door for a future in which the Sanguine is more central than it is now, and in which the contradiction between the business ethics ideal of universal management, leadership, and followership on one hand, and the practical reality of a society divided into managers and non-managers on the other, is less acute than it is now. To that possibility, and to how it might come to pass, we now turn, beginning with another vignette.

Universalizing Business Ethics

In the school district in which I am possibly about to become board president, we wonder and worry about how we are to deal with the sharp inequalities in academic skills—that are manifest from the time children first walk through our doors in kindergarten, and that continue as they grow older and eventually graduate, as a very large majority of our students do, from our high school—and the strong correlation between academic success and race. We also wonder and worry about whether our students are ready to contribute when they move from school to full-time work and how the district's key employees, our teachers, can be leaders, even as they are represented by a union that bargains collectively with us, the board.

My thought is that one key path forward for us in my town and in other communities lies in business education—not the narrowly vocational, old-style, kind, but a new-era version that focuses on ethics, economics, and psychology, on managing people and things. We need to be doing with respect to work what we do now for civics. We need to give children, from their earliest years in school, a sense of themselves as people who will live lives devoted to interacting with other people and material objects in productive enterprises of various kinds, and to managing those relationships. Instead of seeing business ethics as a subject taught mainly to graduate students and undergraduates, we should see it as a subject teachers incorporate into their lesson plans from kindergarten on. For that to happen, we need to give teachers outside the narrow domain of traditionally defined business education the tools and the power to make preparing students for working life part of their calling as elementary school teachers, English teachers, science teachers, art teachers, and so on. We need to open up classrooms to more collaboration, more involvement of people outside the school building, and more focus on human relationships.

All of this, done right, has, I believe, a very great potential to alleviate the painful divides in young people's lives between the academically skilled and the academically less skilled, and in adults' lives between managers and the managed. Starting from the earliest years, universal business ethics and business education—which would embrace nonprofit firms and government agencies as well as for-profit firms—can move us closer to a society in which everyone is a manager, a leader, and a follower. Instead of school being what it is now for many—a repeated experience in being told you are not as good as others—universal business ethics education holds out the promise to everyone, and to the academically less dexterous in particular, of a life in which productive Harmony is achievable. Instead of the transition from education to work being the rude shock that it now is for many of us at all ends of the academic skill spectrum, it would become natural. Lifelong education would become prevalent in a culture in which from our early years we have all been brought up from our earliest years to be simultaneously workers, managers, leaders, and followers.

A Friendly Suggestion for System 2 Elites

A possibility to reflect upon: If management and leadership are to move down the class ladder, it will be partly because we who are System 2-advantaged are no longer as driven as we are now. If we are to make real progress in universalizing management and leadership, perhaps what is most important is for those of us who are at the top of the system now to get out of the way, or to at least open up more room for others. How might that happen? Perhaps we who are the kings and queens of System 2 reason can do better than we have thus far in applying our formidable gifts and skills to Sanguine forms of reason that help us to relax, be happy, and be less compulsive than we are now.[12] As it is, our gifts and skills too often lead us to be more stuck than less System 2-potent people are in unbalanced, rationalized, enduring Melancholy and Choler that keep us ticking, but are not healthy for ourselves, or for others. I have offered ideas in this book as to how we might open the door more to Sanguine reason. I hope for other, and better, ideas from all of us, from all parts of the broad, diverse, multifaceted System 1 and System 2 human spectrums.

A Final Vignette

Thanks in part to thinking about and writing up my reflections on what to tell the board president about my wanting her job, and on comments I got from my undergraduate and graduate business ethics students when I presented an edited version of my issue in my classes, I decided not to give her a line on moving from shared management to shared leadership. Too arrogant, too Choleric, too adversarial. Instead, I told her that I was writing a book on business ethics and leadership, and that I wanted to see how my ideas screwed up in practice, so I could write a sequel. She didn't laugh, but the meeting didn't go too badly. It was a decent version of Phlegmatic Harmony, tinged with Melancholy, I'd say (Figure 6.1).

Ethics	Active/Yang!	Reactive/Yin
Positive	Spirit: Sanguine Type of ethics: **Unclear** Game: Harmony Never ascendant	Spirit: Phlegmatic Type of ethics: **Business Ethics** Game: Imperfect Harmony Ascendant c. 1800–present
Negative	Spirit: Choleric Type of Ethics: **Warrior Ethics** Game: Disharmony Ascendant c. 50,000 BCE–200 BCE	Spirit: Melancholy Type of Ethics: **Priestly Ethics** Game: Partial Disharmony Ascendant c. 200 BCE–1800 CE

Human history over the long haul can be understood in terms of the successive ascendance of different ethics associated with different temperaments, different roles, and different games. Such ethics determine, and are in turn determined by, modes of cultural and economic production. Once, warrior ethics and priestly ethics were ascendant. Now, business ethics is.

Figure 6.1 A History of Ethics

> *Summary*
>
> In the understanding proposed here, business ethics is not an underdog that stands in opposition to ethical squalor, or to an ethical void. Rather, business ethics is a Phlegmatic, practical, pragmatic way of social games that is more productive than, but not morally superior to, other ways of solving social games associated with the other three temperaments. At a broad level, human history can be understood in terms of the successive ascensions of different temperamental ethics, with Phlegmatic business ethics the most recently ascendant form. The rise of business ethics has helped to bring about our current modes of economic and cultural production, and these modes have in turn reinforced the ascendance of business ethics.

> *Exercises*
>
> 1. (a) In addition to Fukuyama's work, discussed in the chapter, there has been a wave of books presenting broad perspectives on human history and modernity in recent years; ten of them are listed and briefly described in a blog post of mine (valuecompetition.typepad.com/value_competition/2012/09/adam-smith-everywhere.html). Using that and other sources, discuss, or reflect on, similarities and differences between those perspectives on history and the "succession of temperamental ethics" perspective presented in the text; explain what you see as the virtues and flaws of the temperamental ethics interpretation of history and modernity compared to other interpretations; (b) Respond to the argument against universal morality and in favor of different temperamental ethics made in the text; (c) Read and discuss, or reflect on, the chapter in *Hard Times* in which Dickens discusses Gradgrind's utilitarian, joyless classroom. Is that a fair picture of the Phlegmatic business ethics temperament?
> 2. (a) Relate the perspective on human history and the possible human future presented here to a science fiction work or works of your choice, or to a popular work of fiction set in the future; (b) Relate the description of the practical business ethics approach to solving social games to a character or characters in popular fiction, the movies, or video games.
> 3. Relate the positions on human moral tribalism taken by Haidt and Greene to the text's case for temperamental ethics. Do you agree with Greene's hope for transcending moral tribalism with liberal utilitarianism, with Haidt's concern about liberalism having its own version of tribalism, or with the text's case for multiple, temperament-based ethics—or with all, or none, of these positions? Why?
> 4. (a) Write your own narrative of social games you play in a given day that are infused with the Phlegmatic, practical spirit of business ethics, as described in the text; (b) Write a narrative of social games you play on a given day that are infused with Sanguine, Choleric, or Melancholy humors; (c) Write your own narrative as to how the scripts in the social games you played on a given day could be flipped or altered to make the games better.

Conclusion

Four Temperaments and four corresponding games govern us, and quite possibly everything else in Nature, if the argument of this book is right. This book has been a preliminary one, which tries to get us to appreciate the division of ourselves into Phlegmatic, Melancholy, Sanguine, and Choleric parts, and the related division of social interactions into four games. If the approach advanced here were to gain currency, future work will rebel, and should rebel, against the simple overarching four-part division I have proposed as a key to understanding our ethical nature. But it is right, I believe, to begin with a simple model.

A central task of this book has been to try to help us take Harmony Games and their central role in our lives seriously. Once we do that, we can perhaps make real progress, not only in the creation of Harmony, the project that I have focused on here, but also in the synthesizing of Harmony and its necessary doppelganger, Disharmony, at work, and elsewhere.

A second central task of the book has been to help us flip standard game-theoretic stories into alternative stories that illustrate the dilemmas of deference, sympathy, and shame, and also illustrate the ways in which we may turn cooler versions of Harmony into happier, more joyous ones. Once we can do that, and can understand the relation of these alternative stories to the standard stories of egoism and competitiveness, we can again perhaps make real progress, not only in this book's project of advancing a new, humanistic, approach to game theory, but also in marrying that new approach and the classical, scientific approach.

Finally, a third central task of the book has been to encourage us to see business ethics not as an embattled, weak underdog, but as a Phlegmatic way of solving social games that is now the world's ascendant ethic. Once we have that understanding, and can appreciate both business ethics and other role ethics that have been ascendant in the past, or that might be in the future, we can, we may hope, enjoy a future in which we have a better balance of the humors, with a rise in the Sanguine element in our lives.

I recently got an email from my sister in Oregon, wondering about subject–verb agreement in the title of this book. Why is it "business ethics matters," rather than "business ethics matter"? Good question! One answer that's fine as far as it goes is that I'm referring to business ethics as a theoretical and practical field, rather than as an individual's (or a firm's) intuitions, emotions, or reasons. But there's another, looser explanation as well, one that gets at the heart of the issues of evolution, meaning, and purpose I've wrestled with throughout this book. Ethics in the four temperaments view espoused here is indeed plural rather than unitary. Business ethics, on the other hand, is a Phlegmatic spirit, not the whole of ethics, much as it is the ascendant form of ethics now. It certainly matters. But so do the different Choleric, Melancholy, and Sanguine ethics that make us up. They all matter. We all matter. Everything that is, was, or will ever be in Nature, including inorganic matter, matters.

Appendices

Appendix A: The Fourfold Classification of Social Games, Related to Schelling's Classification

A Schematic Summary

[From Thomas Schelling, Hockey Helmets, Daylight Saving, and other Binary Choices; the material in brackets and italics consists of my classification of the binary choice games outlined by Schelling into the four categories of Harmony, Imperfect Harmony, Partial Disharmony, and Disharmony.]

It is tempting to work out an exhaustive schematic classification for the various possible binary-choice payoff configurations. But the possibilities, though not endless, are many. The curves, even if monotonic, can be concave or convex, S-shaped, flanged, or tapered; and, of course, they need not be monotonic . . . But with straight lines the number of distinct situations must be limited—at least, the number that is interestingly different. Still, there are at least the following different situations worth distinguishing:

I. There is a unique equilibrium, with all making the same choice.
   It is everybody's favorite outcome. [***Harmony***]
   1. Everybody would be better off if all made the opposite choice.
      a. The collective total would then be at its maximum. [***Disharmony***]
      b. The collective total would be still larger if only some, but not all, made that opposite choice, some then faring better than others but all better than at the equilibrium. [***Disharmony***]
   2. The collective total would be larger if some, not all, made the opposite choice, but some would then be worse off than at equilibrium. [***Disharmony***]
II. There is a unique equilibrium with some choosing L, some R.
   1. All would be better off if all chose R.
      a. The collective total would then be a maximum. [***Partial Disharmony***]
      b. The collective total would be even higher if some still chose L, everybody still being better off than at the equilibrium, but not equally so. [***Partial Disharmony***]
   2. The collective total would be higher, although some people would be worse off, if some (not all) choosing L chose R instead. [***Partial Disharmony***]
   3. The collective total is at a maximum. [***Imperfect Harmony***]

164 • Appendices

III. There are two equilibria, each with all making the same choice.
   1. One of them is everybody's favorite outcome.
      a. The lesser equilibrium, however, is better than most mixtures of choices. [*Imperfect Harmony*]
      b. The lesser equilibrium is worse than most or all mixtures of choices. [*Imperfect Harmony*]
   2. The two equilibria are equally satisfactory and superior to all mixtures of choices. [*Imperfect Harmony*]

Appendix B: Sample Ethical Relations (Harmony) Analysis

You are: Tom Monaghan, CEO, Domino's Pizza
We are: RBS Ethics Consultants, Inc.
Background: You have asked us to analyze how Domino's can best proceed to instill confidence and high performance by major relevant groups in the wake of a decision either (1) to end the company's 30-minute delivery policy or (2) to retain that policy in a modified form stressing safety in the new promotional campaign as well as speed.

What follows is the gist of remarks you (and other Domino's managers) could be making to members of different groups in the event of either decision. After that are brief thoughts on how the analysis could inform your decision about what to do.

	Option A—End the policy	*Option B—Keep a modified policy*
Drivers	[optional—no remarks necessary] "We want you to know we put safety first. Our drivers have an excellent safety record, and we're committed to being the best in the industry. For years, we've had a company-wide bonus pool for locations with 100% safety records. All of our drivers can share in the success of Domino's in being the safest company around." [Phlegmatic Harmony—Practical]	"We're going to be rolling out a new ad campaign with a driver in Ohio who stopped to help an old woman having trouble with her car late at night. The pizza was late, but the customer understood and gave the driver a big tip. At the end of the ad you see the driver and the customer and the old woman all smiling. You guys are our stars!" [Sanguine Harmony—Happy!]
Franchisees	[optional—no remarks necessary] "The decision we've just made was one we thought was important for all of you as well as for the company. We're highly optimistic that our new 'Bring Back the Noid' campaign is going to be our best ever, and we see all of you as vital parts of our success going forward. Here's some more on that . . ." [Phlegmatic Harmony—Practical]	"As you know, our contracts with you make it clear that the liability for accidents rests with the franchise, not with Domino's. At the same time, we want you to understand that we stand behind you 100% going forward. There's lots of legal stuff involved; that's not my department, but I want you know that you have my promise as we roll out our new promotion that Domino's will hold you harmless against the legal sharks for this coming year." [Competitive Harmony—Us–Them]

(*continued*)

(continued)

	Option A—End the policy	Option B—Keep a modified policy
Domino's HQ managers	"This company has been a great love of my life. And as I talk to you now I love it as much as I ever have. It's not my only love—there's Marjorie, there're my daughters, there's my faith. But it's a huge part of me. And you all are, too. You've made Domino's what it is, and I owe you more than I can ever say. I'm committed to Domino's being a great company, with or without me. Someday—not now, not soon, but the day will come—I will be moving on and other people will own the company. That's tough for me to deal with, and it may be for you, too. But it's also healthy. Business and life are changing. We all gotta change, we all gotta grow." [Sanguine Harmony—with accompanying Melancholy]	"We're a different kind of company. Most companies would have caved under the pressure, but we're not going to do it. That has something to do with me. I believe in a higher power beyond all the powers of this earth. That's my business, and I don't impose it on the company or on any of you. But it does have something to do with my absolute commitment that this company will be guided by the highest principles. We will not blow every which way based on fashion or lawsuits. I've hired all of you based on my faith in you as people who are absolutely committed to doing the right thing, and I look forward to moving forward with all of you." [Compliant Harmony—Following]
Potential outside investors	"You Bain guys are the 'unlock the value' people. And you think that if you buy out Domino's from old Tom you can unlock value. That investors will like Domino's better if there isn't a boss that supports right to life and Ave Maria. And maybe you're right. But I can tell you that nobody watches the money more closely than I have and no one is less sentimental. Look at what I did in cutting out our speedy delivery guarantee. This company has been run right. Let me tell you more about that." [Competitive Harmony—Negotiation]	"Look, I don't have to sell. I'm not a guy who gets scared or caves. Look at the way I handled the lawsuit stuff. Not the way a suit would do it. Yes, I'm open to doing a deal with you. But I'm also very good with keeping 100% of Domino's. The number you offer me has to be the right number for me to have any interest." [Competitive Harmony—Negotiation]

Our evaluation: You are likely to get a better response from the drivers, franchisees, and potential investors with Option A and from your managers with Option B.

Our reasoning: Drawing closer to drivers and franchisees with personal, emotion-laden, loyalty-based appeals, as depicted above in Option B, is a risky response to the company's legal troubles. The successful Domino's model has been built on maintaining a measure of relational distance from these groups, and care is called for in changing that model. On the other hand, in regard to the company's current group of Michigan managers, Option B offers a much stronger appeal for loyalty-oriented managers. For such managers, Option A's openness to change is likely to be unsettling, possibly highly so. Finally, for prospective investors, we believe Option A is a signal of investor-oriented management that is much more likely to go along with a good sale price to a private equity firm than Option B is.

Our bottom line: Were Domino's a public corporation, we believe the relational balance would tilt in favor of Option A. Sustaining a tightly knit culture of managerial loyalty, for all its practical and moral value as a consideration in favor of Option B, seems to us less significant than other relational considerations. Given your 100 percent ownership of Domino's, you have a choice as to whether to run your business according to the principles applicable to public corporations or in another fashion; should you want our opinion on that, please contact us.

Appendix C: Sample Prescriptive (Competitive Harmony) Case Analysis

Domino's Pizza ("We'll deliver in 30 minutes" guarantee)

Alternative strategies: (1) Get rid of the guarantee; try to establish other bases on which to differentiate the company's product; (2) Modify the guarantee—for example, establish sliding delivery time targets based on distance, with customers to be informed by cell phone of delays, and/or introduce a new ad campaign emphasizing commitment to safety; (3) Keep the guarantee.

Conclusion

[Of course, you could write a reasoned conclusion the other way.]

Domino's should make a clean break with the past by abandoning a time guarantee for delivering its pizzas. Once, the guarantee policy was a helpful ingredient in Domino's rise from a small outfit to a huge chain. Now, though, the guarantee is about as helpful to Domino's as offering its customers rotten anchovies as a topping.

As a national company with deep pockets that is directly in the sights of aggressive trial lawyers, Domino's is in a different situation from when it was a start-up venture. For a fringe, start-up company, an edgy, ethically tricky approach may be the best one to adopt. But with size comes respectability and responsibility. In terms of the way it should make its corporate decisions, Domino's is now more like Time-Warner than Death Row Records.

As an established pizza chain, Domino's does not want to risk being seen as an aggressive profiteer that encourages its drivers to speed. Domino's is not Ford, FedEx, or UPS, for whom the risks of autos and trucks are understood as inherent in their businesses. When a pizza company like Domino's imposes extra risks of dying on third parties, it creates understandable moral anger. That anger has been turned into jury awards, including one for $78 million in punitive damages,[4] and presents a serious risk to the future of the company.

Jurors' moral intuitions that lawyers rely on to win verdicts may be unfairly tilted against business, and the trial lawyers themselves are anything but saints. But Domino's is a for-profit business. It needs to take the legal system and the moral judgments that

Option 1—End the speed guarantee	Options 2 and 3—Keep or modify the guarantee
1. The rush to deliver is immoral—to gain some extra profit, D is predictably killing people because of employees' incentive to speed.	1. The proposed principle against D's policy is unacceptably broad—speed is a reasonable goal that consumers understandably value.
2. D is a pizza company whose basic job or mission does not involve speed. A rush to deliver by D is morally troubling in a way a FedEx rush is not.	2. D's business model is based on speedy delivery. There is no fundamental moral difference between D and companies like FedEx.
3. Any social value of getting pizzas to people faster is simply not worth the hazards of the policy. D cannot defend killing as many as 20 people per year in order to deliver pizzas a few minutes faster. Any reasonable cost–benefit analysis would show that D's policy—or a revised one with the Internet and cell phones—is inefficient.	3. The cost–benefit calculus may well favor D's policy—for one thing, the critics have no evidence that D's accident rates are worse than for other companies. Twenty deaths (which may be overstated) from 80,000 D's drivers[1] is about the same as the overall US rate of 40,000 deaths[2] from around 200,000,000 drivers.[3]
4. Apart from the other concerns with D's policy, there is an overwhelming practical case for jettisoning it. D needs to take quick action to avoid getting stigmatized as a corporate bad actor, having the juries award huge punitive damages, and possibly having the whole company destroyed.	4. The legal risks are overrated, and in any case it would be wrong for D to allow itself to be stampeded by a media rush to judgment fueled by plaintiff's lawyers. If D doesn't hold the line, it will only lead to worse media frenzies against D and other companies.
5. Clear rules have value: A clear "no guarantee" position is a much more understandable fix for D's problems than an effort to fix the flawed guarantee program with cell phones or a new advertising campaign.	5. Open-ended standards have value: Though there are problems with the way the guarantee has worked, a revised, more nuanced guarantee with advanced technology and a commitment to safety is better than just abandoning the guarantee.
6. D should be very worried about the negative externalities of its guarantee to people hurt by the company: It's D's business to do something, not to wait for outside regulators.	6. D should not be too worried about possible negative externalities, except as they become costs to the company. Let the legal system decide what the costs are.
7. It's moral for the law to impose huge liabilities on D because D's policy encourages speeding and deaths.	7. It's immoral for the law to impose huge liabilities on D because D has tried hard to avoid speeding by its drivers.
8. Intelligent flexibility is a moral virtue: D is not the little company it was; it should reinvent itself by differentiating its product in new ways. Through coming up with new approaches to differentiate the product, D is likely to become a better company.	8. Steadfast dedication is a moral virtue: Though D can be flexible in modifying and updating its guarantee, it should stay dedicated to the policy, partly because people work better when a company is consistent in its values and its long-term strategy.

(continued)

(continued)

Option 1—End the speed guarantee	Options 2 and 3—Keep or modify the guarantee
9. Respect for the basic moral rule that companies like everyone else should respect the law calls for abandoning the guarantee. The point isn't whether the costs of paying out verdicts are small or large; the point is that you have a fundamental moral duty to follow the basic rules laid down in law. Morality cannot be a matter of costs and benefits.	9. D's policy does not violate the law. Negligence law is anything but clear; bad lawyering has lost some cases for D, but good lawyering in others has won. More fundamentally, the right approach is to consider the overall costs and benefits of D's policy given the legal system and all other factors, not to pretend there is a moral rule that solves the issue.
10. Moral Foundations (Haidt): Withdrawing the guarantee is called for under the harm/care foundation—don't hurt innocent people!—and the justice foundation—do the right thing even if it costs you. It makes sense as well under the purity/sacredness foundation, since many people are offended by edgy business conduct like D's. The human gut hates the idea of businesses making money at the expense of human life.	10. Moral Foundations (Haidt): Adhering to the guarantee appeals to people's feelings under the justice foundation—keep your promises and have integrity! It makes sense as well under the loyalty foundation—stick to your group!—and the authority foundation—act like a leader! Also, because the accidents don't involve people in D's care, it does not offend most people's intuition under the harm/care foundation.
11. UMG (Mikhail): The guarantee is condemned by popular opinion for the same reason pushing the fat man to save five people is condemned; D is planning a course of action that sacrifices some people to create benefits for itself and its customers. It's actually worse because D is profiting itself, unlike in the trolley case where the pusher is saving other people's lives.	11. UMG (Mikhail): The guarantee is accepted by popular opinion for the same reason that pulling the switch that kills one person after saving five is acceptable; D is doing its best to serve the public by providing a product speedily, and the harm to a few people is an unwanted side effect of its justified course of action, not the cause of the benefits received by D and its customers.
12. Obedience (Milgram): The basic chilling fact about human nature revealed by Milgram also applies in the D's case. People in a corporate structure like D's will do what they believe the situation demands of them. Drivers operating under a 30-minute guarantee are like Milgram's experimental subjects who pulled the switch. You know that speeding or running a red light is wrong. But faced with a corporate policy and practice that demands delivery by a certain time and with the reality that you can always be let go as a driver if you cost the company money, you will be willing to hurt people as a D's driver.	12. Obedience (Milgram): The Milgram scenario is radically different from the real-world situation in the D's case. Instead of an authoritative researcher telling the subjects to pull the switch, in D's we have independent franchises and drivers making their own choices. The view of drivers as robots speeding to make the 30-minute target is foolish. D's policy is about building corporate good will—"have a pizza on us!"—not about coercing employees. Drivers are likely to get tipped better when they give customers a free or cheap pizza for a slow delivery—given that, the bigger incentive for them is very likely to be safe rather than to speed.

it relies on, including antibusiness judgments, as they are rather than to crusade against them.

Even if it is true, as it may well be, that Domino's drivers cause fewer accidents per pizza delivered than patrons of sit-down pizza parlors cause by their driving, human moral intuition sees the situations as very different. Joshua Greene's and John Mikhail's different approaches to trolley problems converge in helping to explain why Domino's policy, as opposed to an alternative business model that might be associated with more deaths, is likely to disturb people. Per Greene, the "personal/moral" nature of a driver hitting a pedestrian or another car overrides cost–benefit analysis. Per Mikhail, the strategic, profitable nature of the risk created by Domino's speedy delivery policy means that the company will be seen as the cause of harm.

At this point, modifying the guarantee policy by softening it or by instituting an ad campaign stressing the company's commitment to safety isn't the way for Domino's to go. Such a middle-way approach might make sense if Domino's were writing on a blank slate. But it's not. Domino's needs a clear, decisive response to the risk that the guarantee presents.

After ending its guarantee—which should be done simply by stopping the marketing campaign, without any mea culpa that will be used against the company in court—what if Domino's wants at some point in the future to market to nostalgic old customers who remember the old slogan and to prospective customers who value speed?[5] That's fine—but for ethical as well as legal reasons, the company should avoid a 30-minute time guarantee.

Additional Reading

1. Domino's—http://www.fundinguniverse.com/company-histories/domino-s-inc-history/ (corporate history).
2. Tom Monaghan—http://www.epluribusmedia.org/features/2006/0311tom_monaghan.html (activism; sale to Bain).

I highly recommend consulting Wikipedia—I do it all the time—but please refer to sources cited in Wikipedia that you've read, rather than to Wikipedia itself.

Appendix D: The Blame Game

Surveys

[This is the language I used with my executive MBA students in Singapore—as referred to in the text and Figure 5.1, I've also done other versions.]

Version 1

The CEOs of multiple companies are told the following: "We are thinking of building a manufacturing plant in Cambodia. It will help us increase profits no matter what our competitors do, but it will also harm the environment in Cambodia. But if we and our competitors all stay out of Cambodia, we will all do fine, although not quite as well, and the environment in Cambodia will be better." The CEOs all respond that they don't care about harming the environment and just want to make as much profit as possible. The plants are built, profits are made, and the environment is harmed.

Did the CEOs intentionally harm the environment?

Version 2
The CEOs of multiple companies are told the following: "We are thinking of building a manufacturing plant in Cambodia. It will help us increase profits no matter what our competitors do, and it will also help the environment in Cambodia." The CEOs all respond that they don't care about helping the environment and just want to make as much profit as possible. The plants are built, profits are made, and the environment is helped.

Did the CEOs intentionally help the environment?

Version 3
The CEOs of multiple companies are told the following: "We are thinking of building a manufacturing plant in Cambodia. It will help us increase profits no matter what our competitors do, but it will also harm the environment in Cambodia. But if we and our competitors all build plants in Cambodia, the country will be wealthier and its environment will be better." The CEOs all respond that they cannot support building a plant that will harm the environment in Cambodia. No plants are built, and profits are less than they would have been.

Did the CEOs intentionally reduce their companies' profits?

Version 4
The CEOs of multiple companies are told the following: "We are thinking of building a manufacturing plant in Cambodia. It will help us increase profits no matter what our competitors do, but it will also harm the environment in Cambodia. But if we and our competitors all build plants in Cambodia, the country will be wealthier and its environment will be better." The CEOs all respond that they cannot support building a plant that will harm the environment in Cambodia. No plants are built, and Cambodia is poorer and its environment is worse than it would have been if the plants had been built.

Did the CEOs intentionally harm the environment in Cambodia?

Appendix E: Ethical Focal Points—Sample Survey

Instructions: For each of the following eight questions, pick one of the five choices. Imagine as you answer the questions that the following rules apply: A very poor community in a poor part of the world will gain valuable benefits, such as vaccines and pure water, with the size of the benefits depending on how many of you pick the most popular answer (which answer you pick does not matter). So, for example, if there are 48 of you taking the survey and all 48 of you pick the same answer, the community gets 48 units of benefits, while if you are comparatively evenly split, with only 11 of you selecting the most popular answer, the community gets only 11 units of benefits. In other words, the more of you agree on an answer, the better.

1. Pick a number from 1 to 100.
   a. 1;
   b. 7;
   c. 50;
   d. 93;
   e. 100.
2. Pick a place to meet a classmate tomorrow in New York City: You have an important appointment at noon, but no prearranged place to meet, and no way to reach the other person.
   a. The line at the TKTS booth at Times Square;
   b. Under the clock in the main waiting room in Grand Central Station;
   c. By the departure board in the main waiting room at Penn Station;
   d. On the observation deck of the Empire State Building;
   e. The advance ticket sales window at Yankee Stadium.
3. Pick an artist and a work of art.
   a. Eminem, "Stan";
   b. Rembrandt, "The Night Watch";
   c. Lana del Rey, "Ultraviolence";
   d. Katy Perry, "Dark Horse";
   e. Katy Perry, "I Kissed a Girl."
4. Pick an artist and a work of art.
   a. William Shakespeare, "The Merchant of Venice";
   b. Fyodor Dostoevsky, "The Brothers Karamazov";
   c. Britney Spears, "Toxic";
   d. Pablo Picasso, "Guernica";
   e. Jane Austen, "Pride and Prejudice."
5. Pick a person.
   a. Abraham Lincoln;
   b. Josef Stalin;
   c. Kim Il Sung;
   d. Adolf Hitler;
   e. Pol Pot.
6. Pick a person.
   a. Mahatma Gandhi;
   b. Martin Luther King;
   c. Mother Teresa;
   d. Pope John Paul II;
   e. Charles Manson.
7. Pick a game to play.
   a. The Prisoner's Dilemma, in which you and the other player both have to overcome your interests to get to the best result;

b. Harmony, in which and the other player have your interests aligned, so you naturally get to the best result.
  c. The Stag Hunt, in which you and the other player both have to be able to rely on each other to get to the best result;
  d. Chicken, in which you and the other player both have to be able to avoid a temptation to bully if you are to get to the best result;
  e. The Battle of the Sexes, in which one of you has to lead and the other has to follow if you are to get to the best result.
8. Pick a species.
  a. Sea tortoises;
  b. Human beings;
  c. Norway rats;
  d. Koala bears;
  e. Mosquitoes.

Appendix F

Prisoner's Dilemma/Disharmony Exercise

In a few minutes, you will be paired randomly with an anonymous classmate to play a "Prisoner's Dilemma" or "Disharmony" game. The rules of this game are as follows: Both you and the other player will have the choice of playing either C (Cooperate) or D (Defect). If you both play C, you both get 5 extra credit points. If you both play D, you both get 1 extra credit point. If one of you plays D and the other plays C, the one who plays D will get 6 extra credit points, while the one who plays C will get no extra credit points.

Playing D is what is called by game theorists a dominant strategy. No matter whether the other player plays C or D, you get a higher payoff by choosing D rather than C (6 vs. 5 if the other player plays C, 1 vs. 0 if the other player plays D). Playing C is what can be termed a highest joint value strategy, in that it makes it possible that the two of you will receive your highest total possible payoff of 10 extra credit points.

Instructions

1. Sign your name on the paper.
2. Please write FAIR if you think it is fair to play D, and UNFAIR if you think it is unfair.
3. Decide whether you are going to play C or D with the anonymous classmate you will be playing the game with. Write down C or D.
4. Turn in your paper so I can match it up with your anonymous partner.
5. I'll give you your paper back with the score you received, based on what you played and what the other player played. After looking at your score, please write FAIR if you think it is fair to play D, and UNFAIR if you think it is unfair.
6. Turn the paper back in to me.

Appendix G: The Ethical Wisdom of Crowds

Class Exercise

In the space below, circle a number from 1 through 6. The number of points you get depends on what you and your classmates circle. The rules are as follows:

3. If you circle a 4, 5, or 6, you get 3 points if the average of what everyone in the class writes is less than 4, and 1 point if the average of what everyone else writes is 4 or more.
4. If you write down a 1, 2, or 3, you get 2 points if the average of what you and everyone else writes down is less than 4, and zero points if the class average is four or more.

A note on how the game works: No matter what the average number is, you do better personally by circling a 4, 5, or 6, unless your choice pushes the class average to 4 or over. If the average is 4 or over and you circle a 4, 5, or 6, you get 1 point, while if you circle a 1, 2, or 3, you get no points. If the average is under 4 and you circle a 4, 5, or 6, you get 3 points, while if you circle a 1, 2, or 3, you get 3 points. On the other hand, if everyone follows that logic and picks a 4, 5, or 6, all of you get only 1 point. On the other hand, if enough of you pick a 1, 2, or 3, it is possible for all of you to do much better by getting either 2 points or 3 points.

Your number:
1   2   3   4   5   6

Appendix H: Who Leads?

Surveys

Version 1

1. Imagine you and another person in your organization are in a situation in which each of have to make a decision whether to try to lead a project.
   You and the other and the organization will do best overall if one of you leads and the other follows. If you both try to lead or neither of you does, the results will not be good for either of you. Finally, you can decide to not engage in the project at all, in which case the results will also not be good for either of you or the firm.

   You have the following information about the other person: Faced with a challenge from another person, he gets mad in a subtle way. If people challenge him, there will be consequences.

   Taking into account the information about the other person and your own style and approach, what do you do in this situation?

a. I try to lead;
b. I do not try to lead;
c. I may or may not try to lead. It's not exactly random, but it's not clear in advance.
d. I decide not to engage in the project.

Version 2
You have the following information about the other person: Faced with a challenge from another person, he feels good. He believes challenge is good, and he feels good for himself and for the other person.

Version 3
You have the following information about the other person: Faced with a challenge from another person, he feels forgiving and open. He believes that being challenged is part of life, and accepts it.

Version 4
You have the following information about the other person: Faced with a challenge from another person, it doesn't really affect him emotionally one way or another. He is a calm, rational person, who calculates costs and benefits.

Notes

Preface

1. The enjoyment was enhanced greatly by reading Professor Schelling's book, *The Strategy of Conflict*.
2. The key article that helped change the way I thought about law, politics, styles of argument, and eventually game theory was Kennedy (1976).
3. Frank (1988).
4. Eastman (1998).
5. Eastman (1997).
6. The work that helped me the most in doing that was Professor Haidt's 2012 book, *The Righteous Mind*.

Introduction: The Four Temperaments and the Four Games

1. I capitalize "harmony" in the book to reflect the centrality and specialized usage of the concept here. See page 4 for a preliminary explanation of Harmony Games.
2. As with "harmony," I capitalize "phlegmatic," "sanguine," "choleric," "melancholy," and "four temperaments," along with the names of games, to reflect their centrality and specialized usages here.
3. Denby (1965/1979); White (1998).
4. Von Neumann and Morgenstern (1944/2004); Nash (1950); Raiffa and Luce (1958); Schelling (1960/1981).
5. Gintis (2009).
6. McGrath (2014).
7. If you're not familiar with the game, you might want to check out the website: http://www.simcity.com/en_US/manual
8. O'Brien (2013).
9. Rapoport and Chammah (1970).
10. Mathiesen (1999), Grant (2004), and others (such as Professor Schelling in my 1973 class with him) have made this point in regard to the Prisoner's Dilemma. Mathiesen refers to an "Altruist's Dilemma"; this book is devoted in considerable part to trying to persuade us to take that Dilemma seriously and to learn from it.
11. Frank (1988).

12. Maynard Smith and Price (1973) introduced the concept of an evolutionarily stable strategy that can resist incursions by pure egoists in single-shot or repeat games. Axelrod (1984) described the success of a cooperative (but firm) strategy, Tit for Tat, in computer tournaments featuring repeated plays of the Prisoner's Dilemma.

    In regard to business ethics, the present work falls on the strategic, managerial, "ethics (usually) pays" side, as opposed to the religious, "we must be ethical though it may well hurt" side of the dichotomy that Abend (2014) identifies as structuring the field. Paine (2002); Khurana (2007).
13. Arikha (2007).
14. And also in Kant's Konigsberg: In his lectures for his students, published as *Anthropology from a Pragmatic Point of View* (1798/2006), Kant anatomized the classical Four Temperaments. Kant on the Melancholic Temperament: "They do not make promises easily, because they insist on keeping their word, and have to consider whether they will be able to do so."
15. A sense of separation between ethics and nature was, perhaps, especially powerful among intellectuals in the mid to late twentieth century, given the unfortunate politics of the first half of that century—and of World War II and its horrors in particular—that I suggested colored game theory in its original form. Sartre (1938/2000, 1942/1993); Camus (1942a, 1942b).
16. Kahneman (2011). Kahneman's "heuristics and biases" approach tends to emphasize weaknesses in System 1 intuition. Gerd Gigerenzer presents an alternative, more optimistic, "fast and frugal" perspective on intuition (1999); Kelman (2011) reviews the empirical and normative issues at stake in the debate between the two positions. Both Kahneman and Gigerenzer focus on cognitively oriented, low-affect puzzles, rather than on the emotion-laden social games that are the focus of this book. This book's model of the Four Temperaments as devices for the solution of social games could be described as an application, or extension, of Gigerenzer's "fast and frugal" perspective to the domain of ethics, one that is at the same time strongly indebted to Kahneman.
17. Freud (1930/2010).
18. Hume (1739/1985); Haidt (2012).
19. Eysenck (1947/1997); Carducci (2009).
20. The Myers and Briggs Foundation (2015). See also Jung (1971); Keirsey and Bates (1984).
21. Welch and Byrne (2003).
22. Sorkin (2014).
23. Aristotle (1908); Plato (1871); Confucius (1979); Lao Zi (2001).
24. Kant (1783/2001); Bentham (1783/2007).
25. Compare Adam Smith's *Wealth of Nations* (1776/2008), with its steady pulse of calculating utilitarian logic, to his *Theory of Moral Sentiments* (1759/2010). The latter work's anatomizing of socially oriented emotions, brilliant as it often is, does not have the former work's single-minded logic.

26. Such projects of exploring how the world works, and relating that exploration to a conception of lives well lived in business and elsewhere, are flourishing in contemporary business ethics research and teaching in a wide variety of forms, some of which explicitly situate themselves within a virtue-ethics tradition while others do not: Alzola (2012); De Cremer and Tenbrunsel (2012); Hartman (1996); Trevino and Nelson (2011); Paine (2002).
27. Lao Zi (2001).

1   We're Better Than We Think

1. Haidt et al. (1993); Haidt (2001, 2006, 2012); Greene (2013); Boehm (1999, 2012); Henrich et al. (2001); Henrich and Henrich (2007); Bowles and Gintis (2011).
2. "Where Is the Love?," The Black Eyed Peas, 2003.
3. Kant (1783/2001); Bentham (1783/2007).
4. Updike (1972).
5. Camerer (2003).
6. Schelling (1960/1981).
7. *Every Day Is for the Thief* (2014).
8. Perhaps it was this one: *The English Patient* (1992).
9. As an expression of a pessimistic intellectual climate in the 1970s that ffected philosophers as well as psychologists and others, and that drew on the Vietnam War, as well as the factors mentioned in the text, I recommend philosopher Stuart Hampshire's 1973 essay, "Morality and Pessimism."
10. Zimbardo (2008); Kohlberg (1981).
11. Gilligan (1982).
12. Fiske (1992); Shweder (1997).
13. Haidt (2012).
14. Henrich et al. (2001).
15. I would cite the following canonical religious and secular passages as examples of idealistic rigor of different kinds:

> Those who are wise lament neither for the living nor for the dead. Never was there a time when I did not exist, nor you, nor all these kings; nor in the future shall any of us cease to be.
> —Krishna to Arjuna, *The Bhagavad Gita*

> He who does not take the mind and body as "I" and "mine" and who does not grieve for what he has not is indeed called an enlightened one.
> —Buddha, *The Dhammapada*

> Because thou has hearkened unto the voice of thy wife, and hast eaten of the tree, of which I commanded thee, saying, Thou shall not eat of it: cursed is the ground for the sake; in sorrow shall thou eat of it all the days of thy life.
> —*Genesis*, 3:17

> Love your enemies, and pray for those who persecute you, so that you may be sons of your Father who is in heaven . . . For if you love those who love you, what reward have you? Do not even the tax collectors do the same?
> —*The Sermon on the Mount*, Gospel of Matthew

> Woe to those that deal in fraud, those who, when they have to receive by measure from men, exact full measure, but when they have to give by measure or weight to men, give less than due. Do they not think that they will be called to account? On a Mighty Day, a Day when (all) mankind will stand before the Lord of the Worlds?
> —*The Koran*, Sura 83

> I ask myself only: Can you will also that your maxim should become a universal law? If not, then it is reprehensible, and this not for the sake of any disadvantage impending for you or someone else, but because it cannot fit as a principle into a possible universal legislation.
> —Kant, *Prolegomena to the Metaphysics of Morals*

16. Aronson and Budhos (2010).
17. Boehm (1999); Diamond (2012).
18. Goodall (2010); Suddendorf (2003).
19. Wilson (1975/2000).
20. Boehm (1999, 2012); Sober and Wilson (1998); Wilson (2011); Bowles and Gintis (2011). See also Bloom (2013); Kenrick and Griskevicius (2013); Wright (1994).
21. For readers who are interested in the still-intriguing, if no longer as central, debate among Darwinians over group selection for altruistic behavior, as well as in the intricate, indeterminate psychological connections between feelings, ideas, and people, Oren Harman's *The Price of Altruism*, a combined intellectual history and biography of the experimental physicist, evolutionary theorist, unfaithful family man, late-life religious ascetic, and suicide George Price, is very much worth reading.
22. Boehm (1999).
23. Much depends on how low a base number one begins with—starting with two modern humans 130,000 years generates a much more impressive population growth rate for *homo sapiens sapiens* in our very long hunter-gatherer period than starting with 1,000 humans at the same time, and a parallel point applies to the population that left Africa and populated the rest of the world. But in any case, the hunter-gatherer period, in which people gained our major technology—language—and became the planet's dominant land mammal by far, after being a minor, wordless new player at the beginning of the period, should be understand as a period of striking dynamism, like the later agrarian and industrial periods.
24. Singer (1975). For a response to Singer's position that aligns with the position taken in the text on the very high ethical value of the human capability to empathize intuitively with other humans, see Bernard Williams' essay, "The Human Prejudice." Williams (2008). For those of us who like the human voice, lectures

by Williams are available online; one relevant, and in my judgment very good, presentation is available at http://www.youtube.com/watch?v=7p0gAtOTFZg
25. Dawkins (1976/1989).

2   The Harmony Games

1. Foer (2010).
2. Greene (2013) is a good source in favor of the first proposition; Haidt (2012) is a good source in favor of the second.
3. Hu (2011); Davidson (2012).
4. Gilbert (2005).
5. Labich (1994).
6. Robinson (2001); Welch (2003).
7. Edmonds (2007).
8. The claim here is that even writers of works noted for their contentious, aggressive spirit (e.g., Dostoevsky's *Notes from the Underground,* or Nietzsche's *Beyond Good and Evil*) are engaged in Harmonizing with their readers, and likewise with works in a Melancholy spirit (e.g., Krzhizhanovsky's *Autobiography of a Corpse,* Harris's *Fatherland,* and Saramago's *Seeing.*) Such Harmonizing entails conviction on the part of the author that allows readers to join with the author; compare *The Hunger Games,* in which Suzanne Collins amply passes the conviction test, to *Dangerous Games,* a novel with a similar plot, in which Pierre Boule does not believe in his story, resulting in a failure of Harmonizing and an unsuccessful work.

3   Opening the Door to the Sanguine

1. Some nontechnical works of applied game theory that I have been inspired by are Brandenburger and Nalebuff (1996), Chwe (2013), Dixit and Nalebuff (2010). A good text that includes the mainstream game-theoretic stories referred to in this chapter, and many others as well, is Dixit et al. (2009). For perspective on major figures in the field, I recommend Nasar (1998) (John Nash), Nash (2007) (Nash), Dodge (2006) (Schelling), Hendricks and Hansen (2007) (many leading game theorists describe their work), and Harman (2011) (George Price).
2. Knobe (2003).
3. Schelling (1973).
4. Ibid. The four-part classification here also bears some resemblance to Anatol Rapoport's classification of 2 × 2 games as involving four psychological archetypes: exploitation, leadership, heroism, or martyrdom. Rapoport (1967). The psychological archetypes here are different, though: Melancholy yielding (instead of Rapoport's exploitation), Choleric punishment (instead of martyrdom), Phlegmatic rule-following and trust (instead of leadership and heroism), and Sanguine Harmonizing (no parallel). Another source to which I am indebted is Holzinger (2003), which advances a comprehensive classification of 2 × 2 games, with Harmony as one type of game.

5. Widely applied, but not simple in logical or ethical terms; a class on free will and determinism I took with Robert Nozick, and an associated article (1969) in which he posed the conflict between dominance reasoning and expected utility reasoning, has been a background influence on my thinking about game theory over the years.
6. Pinker (2011).
7. Taylor (1911).
8. Mayo (1930/2003); Dickson and Roethlisberger (1939); McGregor (1960/2006).
9. Jensen and Meckling (1976); Jensen and Murphy (1990); Jensen, 2001; Erhard et al. (2009).
10. Sandberg (2013).
11. Jensen (2002).
12. Friedman (1970).
13. Freeman (1984/2010).
14. Carver and Carver (1996).
15. Ratcliff et al. (2012).
16. With its focus on System 2 reasoning about difficult social games in a happy, nonblaming spirit, the approach here is different from, but complementary to, the approaches toward happiness taken by Haidt (2006) and Rubin (2009), which emphasize activities, habits, and worldviews.

4 Bringing Telos Back

1. Dawkins (1976/1989).
2. Wilson (1997).
3. This chapter, as well as Chapter Six, was inspired in part by Robert Wright's *Non-Zero: The Logic of Human Destiny* (2000). Wright's optimistic basic claim that value-adding—or highest joint value, as I call them—outcomes of games tend to prevail over time over value-reducing or value-neutral ones seems to me correct; while Wright fleshes out his claim through historical case studies, I develop my version of the same claim through the Four Temperaments model of how social games are solved by humans and other entities. The model here, though presented in humanistic terms, is capable of being restated in mathematical terms, as noted in the text. My blog, Value Competition, contains many posts in which I explore the ideas that became this book through the use of matrices and the analysis of Nash equilibria.
4. Maynard Smith and Price (1973); Axelrod (1984); Gintis (1999/2009); Bowles and Gintis (2011); Gintis (2009).
5. Degler (1991).
6. Ratcliff et al. (2012).
7. Jekely (2007).
8. Skyrms (2004).
9. Ratcliff et al. (2012).
10. Binmore (1999); Solomon (1999); Grant (2004).
11. Nozick (1969).
12. Bentham (1783/2007).

13. Kant (1783/2001).
14. Buchanan and Tullock (1962/1999).
15. Rawls (1971); Hsieh (2005).
16. Harsanyi (1976).
17. Thanks to Pierre Gagnier for pointing me to Alexander et al., *A Pattern Language* (1977), which focuses on objects and design, and proposes a language of harmonious patterns. It lies beyond the scope of the present work to consider whether the language of social games proposed here is better, or not, than Alexander's pattern language, or other approaches, such as Teilhard de Chardin's vision of evolution to an Omega Point of cosmic unity (1955/2008), or Ray Kurzweil's vision of spiritual machines (1999), that can be understood as attempting to recapture a fused Sanguine and Phlegmatic sense of purpose in nature, and human unity with nature, that one sees in Aristotle and his classical contemporaries.
18. Conan Doyle (1890/2010); von Neumann and Morgenstern (1944/2004).

5 Critical Business Ethics

1. Kennedy (1976); Unger (1975); Weber (1918/2004).
2. Eastman (1996, 1997).
3. Hume (1739/1985); Weber (1918/2004); Kennedy (1976).
4. Williams (2013); Dylan (1967).
5. Hofstadter (1979).
6. Eastman (2013).
7. Weaver and Trevino (1994).
8. Donaldson and Dunfee (1999); Seidman (2007).
9. Two models for rhetorical experimentation: Kennedy and Gabel (1984); Anteby (2013).
10. Williams (1927/1995).
11. Dickson and Roethlisberger (1939); McGregor (1960/2006); Pfeffer (1998).
12. Levitt and Dubner (2005); Levitt and List (2007); Gneezy and List (2013).
13. *Administrative Science Quarterly* (2013).
14. See Appendix B. I've included the argumentative, point–counterpoint prescriptive analysis format, which I also continue to use in my classes, as Appendix C.
15. Monaghan (1986).
16. Pulliam (2003).
17. Walton and Huey (1992); Ortega (2000); Santoro (2000); Chang (2008).
18. Knobe (2003); Appiah (2010).
19. See Appendix D: The Blame Game.
20. *The Strategy of Conflict*. See Appendix E: Ethical Focal Points.
21. A hat tip for this game goes to Rob Kurzban of Penn, who takes an approach to morality rooted in evolutionary psychology (2010), and whom I heard present at NYU. Rob described an experiment in which people were asked to give pre- and post-assessments of the fairness of a 50–50 versus a 72–25 split of earnings from a task. People first got a description of the task, were assigned to different roles in

the task, in which one person does three times as much work as the other, made their fairness assessments, and then were randomly assigned to one of the roles. The results, Rob explained, fit a self-interest pattern—the folks randomly assigned to do only a quarter of the work shifted their assessments in favor of the fairness of the 50–50 split and away from the fairness of the 75–25 split, with the folks randomly assigned to the heavier work condition showing the reverse pattern. That description stimulated me to think of the game described in the text.
22. See Appendix F: The Prisoner's Dilemma/Disharmony.
23. Cialdini (2006); Rosenberg (2011); Brooks (2011).
24. Surowiecki (2004).
25. See Appendix G: The Ethical Wisdom of Crowds.
26. Colin Camerer's (2003) *Behavioral Game Theory* is an excellent source on the experimental literature. See also Smith (2003); Katok (2010).
27. Schelling (1960/1981). See Appendix H: Who Leads?
28. Pulliam (2003).

6 Why Business Ethics Matters

1. Greene (2013); Haidt (2012). Under Moral Foundations Theory as developed by Haidt and his collaborators, those of us who are more liberal, highly educated, and modern are not better, but are simply different, and narrower, in our morality. To use the phrase coined by Henrich et al. (2010), we are WEIRD—Western, Educated, Industrial, Rich, and Democratic. Haidt's studies show that both the WEIRD and the non-WEIRD value the moral foundations of freedom, fairness, and care highly. But while the moral taste buds of working-class, conservative, and developing nation respondents are also highly attracted by purity and repelled by disgusting acts, disrespect for authority, and disloyalty, the WEIRD among us are not similarly morally repelled, and do not value the moral foundations of sanctity, authority, and loyalty as highly as our non-WEIRD peers. For his part, Greene relies on studies he has conducted (Greene et al., 2001) showing that emotion-oriented parts of the brain light up in functional MRI scans when people make a judgment not to push a man to his death, while calculation and cognition-oriented centers are activated when they make the utilitarian judgment that pushing the man to save five lives is acceptable. In a doubling-down on what Henrich and Haidt might call WEIRD morality, Greene argues that we should be skeptical of the intuitive judgment that pushing the man is repugnant. The right morality, Greene argues in the conclusion to *Moral Tribes*, is a liberal, universalistic utilitarianism that eschews parochial moral communities, and that is capable of overriding intuitive repugnance to achieve the greater social good. Weaver and Brown (2012).
2. Hegel (1830/1956).
3. Engels (1884).
4. Ibid.

5. Marx and Engels (1848).
6. Spencer (1884).
7. Boehm (1999).
8. Lanier (2013).
9. Schelling, 1978.
10. James (1907/1995); Dewey (1929/2013); Rorty (1979); Fish (1980); Frank (1930/1970); Kennedy (1997); Unger (1987).
11. Rose (2011).
12. Skidelsky and Skidelsky (2012).

Appendices

1. http://findarticles.com/p/articles/mi_m3190/is_n33_v23/ai_7865517/ (1989; contains 20 Domino's accident deaths and 80,000 drivers figures).
2. http://www.infoplease.com/ipa/A0908129.html (US driving deaths).
3. http://www.statemaster.com/graph/trn_lic_dri_tot_num-transportation-licensed-drivers-total-number (number of US drivers).
4. http://www.snopes.com/business/consumer/dominos.asp (lawsuits that motivated Domino's to drop its guarantee).
5. http://blogs.wsj.com/law/2007/12/17/dominos-pizza-amp-the-law/ (Domino's new 30-minute "non-guarantee").

References

Abend, Gabriel. *The Moral Background*. Princeton: Princeton University Press, 2014.
Alexander, Christopher, Sara Ishikawa, and Murray Silverstein. *A Pattern Language: Towns, Buildings, Construction*. New York: Oxford University Press, 1977.
Alzola, Miguel. The possibility of virtue, *Business Ethics Quarterly*, 22 (2012): 377–404.
Anteby, Michel. *Manufacturing Morals*. Chicago: University of Chicago Press, 2013.
Appiah, Kwame Anthony. *Experiments in Ethics*. Cambridge: Harvard University Press, 2010.
Arikha, Noga. *Passions and Tempers*. New York: HarperCollins, 2007.
Aristotle. *Physics*. Translated by R. P. Hardie and R. K. Gaye. The Internet Classics Archive, 1930, http://classics.mit.edu/Aristotle/physics.html
Aristotle. *The Nicomachean Ethics*. Translated by W. D. Ross. The Internet Classics Archive, 1908, http://classics.mit.edu/Aristotle/nicomachaen.htm
Aronson, Marc and Marina Budhos. *Sugar Changed the World: A Story of Magic, Spice, Slavery, Freedom, and Science*. New York: Clarion Books, 2010.
Austen, Jane. *Pride and Prejudice*, 1813, http://www.mollands.net/etexts/prideandprejudice/pnp14.htm
Axelrod, Robert. *The Evolution of Cooperation*. New York: Basic Books, 1984.
Bentham, Jeremy. *Principles of Morals and Legislation*. Mineola: Dover Publications, 1783/2007.
Berne, Eric. *Games People Play*. New York: Random House, 1964/1996.
Bhagavad Gita. http://www.bartleby.com/45/4/2.html
Binmore, Ken. Game theory and business ethics, *Business Ethics Quarterly*, 9, no. 1 (1999): 31–35.
Bloom, Paul. *Just Babies*. New York: Random House, 2013.
Bob Dylan and The Band. Tears of rage, *Columbia Records*, 1967.
Boehm, Christopher. *Hierarchy in the Forest*. Cambridge: Harvard University Press, 1999.
Boehm, Christopher. *Moral Origins*. New York: Basic Books, 2012.
Boule, Pierre. *Dangerous Games*. London: Hesperus Press Limited, 2014.
Bowles, Samuel and Herbert Gintis. *A Cooperative Species: Human Reciprocity and Its Evolution*. Princeton: Princeton University Press, 2011.
Brandenburger, Adam and Barry Nalebuff. *Co-opetition*. New York: Doubleday, 1996.

Brooks, David. *The Social Animal: The Hidden Sources of Love, Character, and Achievement.* New York: Random House, 2011.

Bryan, Nicole, Wayne Eastman, Sasha Poucki, and Anne Quarshie. Bringing human trafficking into executive suites and academia (Paper presented at the *Academy of Management* annual conference, 2014).

Buchanan, James M. and Gordon Tullock. *The Calculus of Consent.* Indianapolis, IN: Liberty Fund, 1962/1999.

Cable, Daniel, Francesca Gino, and Bradley Staats. Breaking them in or eliciting their best? Reframing socialization around newcomers' authentic self-expression, *Administrative Science Quarterly*, 58 (2013): 1–36.

Camerer, Colin. *Behavioral Game Theory: Experiments in Strategic Interaction.* Princeton: Princeton University Press, 2003.

Camus, Albert. *The Stranger.* New York: Alfred A. Knopf, 1942a.

Camus, Albert. *The Myth of Sisyphus.* Paris: Gallimard, 1942b.

Carducci, Bernardo. *The Psychology of Personality: Viewpoints, Research, and Applications.* Hoboken: John Wiley & Sons, 2009.

Carver, John and Miriam Mayhew Carver. *Basic Principles of Policy Governance.* The CarverGuide Series on Effective Board Governance, no. 1. San Francisco: Jossey-Bass, 1996.

Chang, Leslie. *Factory Girls.* New York: Random House, 2008.

Chwe, Michael Suk-Young. *Jane Austen, Game Theorist.* Princeton: Princeton University Press, 2013.

Cialdini, Robert. *Influence: The Psychology of Persuasion.* New York: Harper Business, rev. ed., 2006.

Cole, Teju. *Every Day is for the Thief.* New York: Random House, 2014.

Collins, Suzanne. *The Hunger Games.* New York: Scholastic Press, 2008.

Conan Doyle, Arthur. The final problem. In *The Memoirs of Sherlock Holmes.* Mineola: Dover Books, 1890/2010, p. 185, https://en.wikisource.org/wiki/The_Final_Problem

Confucius. *The Analects.* Translated by D. C. Lau. New York: Penguin Books, 1979.

Darwin, Charles. *The Descent of Man.* Seattle: Pacific Publishing, 2011.

Davidson, Amy. The Tyler Clementi verdict, *The New Yorker*, March 16, 2012.

Dawkins, Richard. *The River out of Eden.* London: Basic Books, 1996.

Dawkins, Richard. *The Selfish Gene.* Oxford: Oxford University Press, 1976/1989.

De Cremer, David and Ann E. Tenbrunsel (eds.). *Behavioral Business Ethics: Shaping an Emerging Field.* New York: Routledge, 2012.

Degler, Carl. *In Search of Human Nature: The Decline and Revival of Darwinism in American Social Thought.* New York: Oxford University Press, 1991.

Denby, Edwin. *Dancers, Buildings and People in the Streets.* New York: Popular Library, 1965/1979.

Dewey, John. *The Quest for Certainty.* New York: Isha Books, 1929/2013.

Diamond, Jared. *The World Until Yesterday: What We Can Learn from Traditional Societies.* New York: Viking Penguin, 2012.

Dickson, William and Harold Roethlisberger. *Management and the Worker: An Account of the Research Program Conducted by the Western Electric Company, Hawthorne Works, Chicago.* Cambridge: Harvard University Press, 1939.

Dixit, Avishai and Barry Nalebuff. *The Art of Strategy*. New York: W. W. Norton, 2010.

Dixit, Avishai, Susan Skeath, and David H. Reiley, Jr. *Games of Strategy* (3rd ed.). New York: W. W. Norton, 2009.

Dodge, Robert. *The Strategist: The Life and Times of Thomas Schelling*. Cambridge, MA: Hollis, 2006.

Donaldson, Thomas and Thomas W. Dunfee. *Ties That Bind: A Social Contracts Approach to Business Ethics*. Boston: Harvard Business School Press, 1999.

Donne, John. The first anniversary, 1611, www.luminarium.org/renascence-editions/donne1.html

Dostoevsky, Fyodor. *Notes from the Underground*. New York: Tribeca Books, 2010.

Dostoevsky, Fyodor. *The Brothers Karamazov*. Translated by Richard Pevear and Larissa Volokhonsky. New York: Farrar, Straus and Giroux, 2002.

Drucker, P. F. What is business ethics?, *The Public Interest*, no. 63 (1981): 18–36.

Eastman, Wayne. "Everything's up for grabs": The Coasean story in game-theoretic terms, *New England Law Review*, 31 (1996): 1–37.

Eastman, Wayne. Ideology as rationalization and as self-righteousness: Psychology and law as paths to critical business ethics, *Business Ethics Quarterly*, 23 (2013): 527–560.

Eastman, Wayne. Telling alternative stories: Heterodox versions of the Prisoner's Dilemma, the Coase Theorem, and supply-demand equilibrium, *Connecticut Law Review*, 29 (1997): 727–805.

Eastman, Wayne. Working for position: Women, men, and managerial work hours, *Industrial Relations*, 37 (1998): 51–56.

Eastman, Wayne and Michael Santoro. The importance of value diversity in corporate life, *Business Ethics Quarterly*, 13 (2003): 433–452.

Edmonds, David. *Rousseau's Dog: Two Great Thinkers at War in the Age of Enlightenment*. New York: Harper Perennial, 2007.

Engels, Friedrich. *The Origin of the Family, Private Property, and the State*, 1884, www.marxists.org/archive/marx/works/1884/origin-family/

Erhard, Werner, Michael C. Jensen, and Steve Zaffron. Integrity: A positive model that incorporates the normative phenomena of morality, ethics, and legality, 2009, www.papers.ssrn.com/sol3/papers.cfm?abstract_id=920625

Eysenck, Hans. *Dimensions of Personality*. New Brunswick, NJ: Transaction, 1947/1997.

Fish, Stanley. *Is There a Text in This Class?* Cambridge, MA: Harvard University Press, 1980.

Fiske, Alan. The four elementary forms of sociality: Framework for a unified theory of social relations, *Psychological Review*, 99 (1992): 689–723.

Foer, Franklin. *How Soccer Explains the World: An Unlikely Theory of Globalization*. New York: HarperCollins, 2010.

Frank, Jerome. *Law and the Modern Mind*. New Brunswick, NJ: Transaction, 1930/1970.

Frank, Robert. *Passions within Reason: The Strategic Role of the Emotions*. New York: W. W. Norton, 1988.

Frank, Robert. *What Price the Moral High Ground?*. Princeton: Princeton University Press, 2004.

Franklin, Benjamin. *Poor Richard's Almanac and Other Writings*. Mineola: Dover Publications, 2013.

Freeman, Edward. *Strategic Management: A Stakeholder Approach*. Cambridge: Cambridge University Press, 1984/2010.

Freud, Sigmund. *Civilization and Its Discontents*. Mansfield Center: Martino Publishing, 1930/2010.

Friedman, Milton. The social responsibility of business is to maximize its profits, *The New York Times Magazine*, September 13, 1970.

Fukuyama, Francis. The end of history?, *The National Interest*, 1989.

Fukuyama, Francis. *The End of History and the Last Man*. New York: Avon Books, 1992.

Gigerenzer, Gerd, Peter M. Todd, and the ABC Research Group. *Simple Heuristics That Make Us Smart*. New York: Oxford University Press, 1999.

Gilbert, Daniel. *Stumbling on Happiness*. New York: Random House, 2005.

Gilligan, Carol. *In a Different Voice*. Cambridge, MA: Harvard University Press, 1982.

Gintis, Herbert. *Game Theory Evolving: A Problem-Centered Introduction to Modeling Strategic Interaction* (2nd ed.). Princeton: Princeton University Press, 1999/2009.

Gintis, Herbert. *The bounds of reason: Game theory and the unification of the behavioral sciences*. Princeton, NJ: Princeton University Press, 2009.

Gneezy, Uri and John List. *The Why Axis: Hidden Motives and the Undiscovered Economics of Everyday Life*. New York: PublicAffairs, 2013.

Goodall, Jane. *Through a Window: My Thirty Years with the Chimpanzees of Gombe*. New York: Houghton Mifflin Harcourt, 2010.

Goodwin, Doris Kearns. *Team of Rivals: The Political Genius of Abraham Lincoln*. New York: Simon and Schuster, 2005.

Graeber, David. *Debt: The First 5,000 Years*. New York: Melville House Publishing, 2011.

Grant, Colin. The altruists' dilemma. *Business Ethics Quarterly*, 14 (2004): 315–328.

Greene, J. D., R. B. Sommerville, L. E. Nystrom, J. M. Darley, and J. D. Cohen. An fMRI investigation of emotional engagement in moral judgment, *Science*, 293 (2001): 2105–2108.

Greene, Joshua. *Moral Tribes: Emotion, Reason, and the Gap Between Us and Them*. New York: Penguin Press, 2013.

Haidt, J., S. Koller, M. and Dias. Affect, culture, and morality, or is it wrong to eat your dog?, *Journal of Personality and Social Psychology*, 65 (1993): 613–628.

Haidt, Jonathan. The bright future of post-partisan social psychology (Presentation, *Society for Personality and Social Psychology*, San Antonio, TX, January 27, 2011). Accessed on December 14, 2014, http://people.stern.nyu.edu/jhaidt/postpartisan.html

Haidt, Jonathan. The emotional dog and its rational tail: A social intuitionist approach to moral judgment, *Psychological Review*, 108 (2001): 814–834.

Haidt, Jonathan. *The Happiness Hypothesis: Finding Modern Truth in Ancient Wisdom*. New York: Basic Books, 2006.

Haidt, Jonathan. *The Righteous Mind: Why Good People Are Divided by Politics and Religion*. New York: Random House, 2012.

Halley, Janet. *Split Decisions: How and Why to Take a Break from Feminism*. Princeton: Princeton University Press, 2006.

Hampshire, Stuart. Morality and pessimism, *The New York Review of Books*, January 1973.

Harman, Oren. *The Price of Altruism: George Price and the Search for the Origins of Kindness*. New York: W. W. Norton, 2011.

Harris, Richard. *Fatherland: A Novel*. New York: HarperCollins, 2006.

Harsanyi, John C. Can the maximin principle serve as a basis for morality? A critique of John Rawls's theory, *American Political Science Review*, 69 (1975): 594–606.

Hartman, Edwin M. *Organizational Ethics and the Good Life*. Ruffin Series in Business Ethics. New York: Oxford University Press, 1996.

Heath, Joseph. Business ethics without stakeholders, *Business Ethics Quarterly*, 16 (2006): 533–557.

Hegel, G. W. F. *Lectures on the Philosophy of History*. Translated by J. Sibree. Mineola: Dover Publications, 1830/1956.

Hegel, G. W. F. *The Phenomenology of Spirit*. Translated by A. V. Miller. New York: Oxford University Press, 1807/1976.

Hendricks, Vincent F. and Pelle G. Hansen (eds.). *Game Theory: 5 Questions*. LaVergne, TN: Automatic Press, 2007.

Henrich, Joseph, et al. In search of Homo Economicus: Behavioral experiments in 15 small-scale societies, *American Economic Review*, 92 (2001): 73–78.

Henrich, Joseph and Natalie Henrich. *Why Humans Cooperate: A Cultural and Evolutionary Explanation*. New York: Oxford University Press, 2007.

Henrich, Joseph, Steven Heine, and Ara Norenzayan. The weirdest people in the world?, *Behavioral and Brain Sciences*, 33 (2010): 61–135.

Hindemith, Paul. *The Four Temperaments*. Score to a ballet choreographed by George Balanchine, November 1946.

Hobbes, Thomas. *Leviathan*. Cambridge: Cambridge University Press, 1651/1996.

Hofstadter, Douglas. *Godel, Escher, Bach: An Eternal Golden Braid*. New York: Basic Books, 1979.

Holzinger, Katharina. The problems of collective action: A new approach, 2003, www.coll.mpg.de/pdf_dat/2003_02online.pdf

Hsieh, Nien-he. Rawlsian justice and workplace republicanism, *Social theory and Practice*, 31 (2005): 115–142.

Hu, Winnie. Bullying law puts New Jersey schools on spot, *The New York Times*, August 2011. Accessed on December 15, 2014, http://www.nytimes.com/2011/08/31/nyregion/bullying-law-puts-new-jersey-schools-on-spot.html?pagewanted=all&_r=0

Hume, David. *A Treatise of Human Nature*. New York: Penguin Books, 1739/1985.

Huntington, Samuel. *The Clash of Civilizations and New World Order*. New York: Simon & Schuster, 1996.

James, Henry. *The Tragic Muse*. Charleston: BiblioLife, 1890/2009.

James, William. *Pragmatism*. Charleston: BiblioLife, 1907/1995.

Jekely, Gaspar. Origin of phagotrophic eukaryotes as social cheaters in microbial biofilms, *Biology Direct*, 2 (2007): 1–3.

Jensen, Michael C. Corporate budgeting is broken, let's fix it, *Harvard Business Review*, November (2001): 94–101.

Jensen, Michael C. Value maximization, stakeholder theory, and the corporate objective function, *Business Ethics Quarterly*, 12 (2002): 235–256.
Jensen, Michael C. and Kevin Murphy. CEO incentives—It's not how much you pay, but how. *Harvard Business Review*, May–June (1990): 138–153.
Jensen, Michael C. and William H. Meckling. Theory of the firm: Managerial behavior, agency costs, and ownership structure, *Journal of Financial Economics*, 3 (1976): 305–360.
Jung, Carl Gustav. *Psychological Types* (R. F. C. Hull, ed.). Princeton: Princeton University Press, 1971.
Kahneman, Daniel. *Thinking Fast and Slow*. New York: Farrar, Straus and Giroux, 2011.
Kant, Immanuel. *Anthropology from a Pragmatic Point of View*. Cambridge: Cambridge University Press, Cambridge Texts in the History of Philosophy, 1798/2006.
Kant, Immanuel. *Critique of Pure Reason*. Translated by Marcus Weigelt. London: Penguin Books, 1781/2007.
Kant, Immanuel. *Prolegomena to the Metaphysics of Morals*. Translated by James Ellington. Indianapolis: Hackett Publishing Company, 1783/2001.
Katok, Elena. Using laboratory experiments to build better operations management models, *Foundations and Trends in Technology, Information and Operations Management*, 5 (2010): 1–84.
Keirsey, David and Marilyn Bates. *Please Understand Me: Character and Temperament Types*. Del Mar: Prometheus Nemesis Book Company, 1984.
Kelman, Mark. *The Heuristics Debate*. Oxford: Oxford University Press, 2011.
Kelman, Mark. Trashing, *Stanford Law Review*, 36 (1984): 293–348.
Kennedy, Duncan M. *A Critique of Adjudication [fin de siecle]*. Cambridge: Harvard University Press, 1997.
Kennedy, Duncan M. Form and substance in private law adjudication, *Harvard Law Review*, 89 (1976): 1685.
Kennedy, Duncan M. The structure of Blackstone's Commentaries, *Buffalo Law Review*, 28 (1979): 209–382.
Kennedy, Duncan M. and Peter Gabel. Roll over Beethoven, *Stanford Law Review*, 36 (1984): 1–55.
Kenrick, Douglas and Vladas Griskevicius. *The Rational Animal: How Evolution Made Us Smarter Than We Think*. New York: Basic Books, 2013.
Khurana, Rakesh. *From Higher Aims to Hired Hands: The Social Transformation of American Business Schools and the Unfulfilled Promise of Management as a Profession*. Princeton: Princeton University Press, 2007.
Knobe, Joshua. Intentional action and side effects in ordinary language, *Analysis*, 63 (2003): 190–193.
Kohlberg, Lawrence. *The Philosophy of Moral Development: Moral Stages and the Idea of Justice*. New York: Harper and Row, 1981.
Krzhizhanovksy, Sigizmund. *The Autobiography of a Corpse*. New York: NYRB Classics, 2013.
Kurzban, Robert. *Why Everyone (Else) Is a Hypocrite: Evolution and the Modular Mind*. Princeton: Princeton University Press, 2010.

Kurzweil, Ray. *The Age of Spiritual Machines: When Computers Exceed Human Intelligence.* New York: Penguin Books, 1999.

Labich, Kenneth. Is Herb Kelleher America's best CEO? *Fortune,* May 1994. Accessed in June 2015, archive.fortune.com/magazines/fortune/fortune_archive/1994/05/02/79246/index.htm

Lanier, Jaron. *Who Owns the Future?.* New York: Simon & Schuster, 2013.

Lao Zi, *Dao-de-Jing.* London: University of California Press, 2001.

Leaf, Munro and Robert Lawson. *The Story of Ferdinand.* New York: Puffin Books, 2011.

Levitt, Steven and Stephen Dubner. *Freakonomics: A Rogue Economist Explores the Hidden Side of Everything.* New York: William Morrow, 2005.

Levitt, Steven D. and John A. List. What do laboratory experiments measuring social preferences reveal about the real world?, *Journal of Economic Perspectives,* 21 (2007): 153–174.

MacIntyre, Alasdair. *After Virtue.* Notre Dame: Notre Dame Press, 1981/2007.

Malik, Kenan. *The Quest for a Moral Compass: A Global History of Ethics.* New York: Melville House Publishing, 2014.

Marx, Karl. *Economic and Philosophical Manuscripts of 1844,* 1844, https://www.marxists.org/archive/marx/works/1844/manuscripts/needs.htm

Marx, Karl and Friedrich Engels. *The Communist Manifesto,* 1848, https://www.marxists.org/archive/marx/works/1848/communist-manifesto/

Marx, Karl. *The Poverty of Philosophy,* 1847, https://www.marxists.org/archive/marx/works/1847/poverty-philosophy/index.htm

Mathiesen, Kay. Game theory in business ethics: Bad ideology or bad press?, *Business Ethics Quarterly,* 9 (1999): 37–45.

Maynard Smith, John and George R. Price. The logic of animal conflict, *Nature,* 246 (1973): 15–18.

Mayo, Elton. *The Human Problems of an Industrial Civilization.* New York: Routledge, 1930/2003.

McGrath, Ben. Good game: The rise of the professional cyber athlete, *The New Yorker,* November 24, 2014.

McGregor, Douglas. *The Human Side of Enterprise.* New York: McGraw-Hill, 1960/2006.

Molinsky, Andrew and Joshua D. Margolis. Necessary evils and interpersonal sensitivity in organizations, *Academy of Management Review,* 30 (2005): 245–268.

Monaghan, Thomas. *Pizza Tiger.* New York: Random House, 1986.

Nasar, Sylvia. *A Beautiful Mind.* New York: Simon and Schuster, 1998.

Nash, John. Equilibrium points in n-person games, *Proceedings of the National Academy of Sciences of the United States of America,* 36 (1950): 48–49.

Nash, John. *The Essential John Nash* (Harold W. Kuhn and Sylvia Nasar, eds.). Princeton, NJ: Princeton University Press, 2007.

Nietzsche, Friedrich. *Beyond Good and Evil.* Translated by R. J. Hollingdale. New York: Penguin Books, 2003.

Nietzsche, Friedrich. *The Genealogy of Morals.* Mineola: Dover Publications, 2003.

Norman, Wayne. Business ethics as self-regulation, *Journal of Business Ethics,* 102 (2011): 43–57.

Nozick, Robert. Newcomb's problem and two principles of choice. In Nicholas Rescher (ed.), *Essays in Honor of Carl G Hempel*. Amsterdam: Springer, 1969.
O'Brien, Stephen. *The Ultimate Player's Guide to Minecraft*. New York: Que, 2013.
Ondaatje, Michael. *The English Patient*. New York: Vintage, 1992.
Ortega, Bob. *In Sam We Trust: The Untold Story of Sam Walton and Wal-Mart, the World's Most Powerful Retailer*. New York: Three Rivers Press, 2000.
Paine, Lynn Sharp. *Value Shift: Why Companies Must Merge Social and Financial Imperatives to Achieve Superior Performance*. New York: McGraw-Hill, 2002.
Pascal, Blaise. *Pensees*. Translated by A. J. Krailsheimer. New York: Penguin Books, 1995.
Pfeffer, Jeffrey. *The Human Equation: Building Profits by Putting People First*. Boston: Harvard Business School Press, 1998.
Pinker, Steven. *The Better Angels of Our Nature*. New York: Penguin Books, 2011.
Plato. *The Republic*. Translated by B. Jowett, 1871, http://classics.mit.edu/Plato/republic.html
Pulliam, Susan. Ordered to commit fraud, a staffer balked, then caved, *The Wall Street Journal*, June 2003. Accessed on December 15, 2014, http://www.wsj.com/articles/SB105631811322355600
Raiffa, Duncan and Howard Luce. *Games and Decisions: Introduction & Critical Survey*. New York: John Wiley & Sons, 1958.
Rapoport, Anatol. Exploiter, leader, hero, and martyr: The four archetypes of the 2 x 2 game, *Behavioral Science*, 12 (1967): 81–84.
Rapoport, Anatol. *Two-Person Game Theory: The Essential Ideas*. Ann Arbor: University of Michigan Press, 1966.
Rapoport, Anatol and Albert Chammah. *The Prisoner's Dilemma*. Ann Arbor: University of Michigan Press, 1970.
Ratcliff, W. C., R. F. Denison, M. Borrello, and M. Travisano. Experimental evolution of multicellularity, *Proceedings of the National Academy of Sciences of the United States of America*, 109 (2012): 1595–1600.
Rawls, John. *A Theory of Justice*. Cambridge: Harvard University Press, 1971.
Robinson, James. *Jack Welch on Leadership: Executive Lessons from the Master CEO*. Roseville: Prima Lifestyles, 2001.
Rorty, Richard. *Philosophy and the Mirror of Nature*. Princeton: Princeton University Press, 1979.
Rose, David. *The Moral Foundations of Economic Behavior*. Oxford: Oxford University Press, 2011.
Rosenberg, Tina. *Join the Club: How Peer Pressure Can Transform the World*. New York: W. W. Norton, 2011.
Rubin, Gretchen. *The Happiness Project*. New York: HarperCollins, 2009.
Russell, Bertrand. *The History of Western Philosophy*. New York: Simon & Schuster, 1945/1972.
Sandberg, Sheryl. *Lean In: Women, Work, and the Will to Lead*. New York: Knopf, 2013.
Santoro, Michael. *Profits and Principles: Global Capitalism and Human Rights in China*. Ithaca: Cornell University Press, 2000.
Saramago, Jose. *Seeing*. Orlando: Harcourt Books, 2004.

Sartre, Jean-Paul. *Being and Nothingness*. New York: Washington Square Press, 1942/1993.
Sartre, Jean-Paul. *Nausea*. London: Penguin Modern Classics, 1938/2000.
Schelling, Thomas. Egonomics, or the art of self-management, *American Economic Review*, 68 (1978): 290–294.
Schelling, Thomas. Ethics, law, and the exercise of self-command. In John Rawls and Sterling M. McMurrin (eds.), *Liberty, Equality, and Law: Selected Tanner Lectures on Moral Philosophy*. Salt Lake City: University of Utah Press, 1987.
Schelling, Thomas. Hockey helmets, concealed weapons, and daylight saving: A study of binary choices with externalities, *The Journal of Conflict Resolution*, 17 (1973): 321–428.
Schelling, Thomas. *Micromotives and Macrobehavior*. New York: W. W. Norton, 2006.
Schelling, Thomas. Strategy and self-command, *Negotiation Journal*, 5 (1989): 343–347.
Schelling, Thomas. *The Strategy of Conflict*. Cambridge: Harvard University Press, 1960/1981.
Seidman, Dov. *How: Why How We Do Anything Means Everything*. New York: Wiley, 2007.
Sen, Amartya. Does business ethics make economic sense?, *Business Ethics Quarterly*, 3 (1993): 45–54.
Shweder, Richard A., Nancy C. Much, Manamohan Mahapatra, and Larry Park. The "big three" of morality (autonomy, community, divinity), and the "big three" explanations of suffering. In Allan Brandt and Paul Rozin (eds.), *Morality and Health*, 119–169. New York: Routledge, 1997.
Sim City official site: http://www.simcity.com/en_US/manual
Singer, Peter. *Animal Liberation*. New York: Avon Books, 1975.
Skidelsky, Robert and Edward Skidelsky. *How Much Is Enough?*. New York: Other Press, 2012.
Skyrms, Brian. *The Stag Hunt and the Evolution of the Social Structure*. Cambridge: Cambridge University Press, 2004.
Smith, Adam. *The Theory of Moral Sentiments*. New York: Digireads, 1759/2010.
Smith, Adam. *The Wealth of Nations*. New York: Oxford University Press, 1776/2008.
Smith, Vernon L. Constructivist and ecological rationality in economics, *American Economic Review*, 93 (2003): 465–508.
Sober, Elliott and David Sloan Wilson. *Unto Others: The Evolution and Psychology of Unselfish Behavior*. Cambridge MA: Harvard University Press, 1998.
Solomon, Robert. Game theory as a model for business and business ethics, *Business Ethics Quarterly*, 9 (1999): 11–29.
Solomon, Robert. *True to Our Feelings*. Oxford: Oxford University Press, 2001.
Solomon, Robert. Victims of circumstances? A defense of virtue ethics in business, *Business Ethics Quarterly*, 14 (2003): 43–62.
Sorkin, Andrew Ross. Berkshire's Radical Strategy: Trust, *The New York Times*, May 2014. Accessed December 14, 2014, http://dealbook.nytimes.com/2014/05/05/berkshires-radical-strategy-trust/
Spencer, Herbert. *The Man versus the State*, 1884, www.econlib.org/library/LFBooks/Spencer/spnMvS0.html

Suddendorf, Thomas. *The Gap: The Science of What Separates Us from Other Animals.* New York: Basic Books, 2003.
Surowiecki, James. *The Wisdom of Crowds.* New York: Random House, 2004.
Taylor, Frederick Winslow. *The Principles of Scientific Management.* New York: Harper & Brothers, 1911.
Teilhard de Chardin, Pierre. *The Phenomenon of Man.* New York: Harper, 1955/2008.
The Black-Eyed Peas and Justin Timberlake. Where is the love?, *A&M Records*, 2003.
The Dhammapada. Translated by Eknath Easwaran. Berkeley: The Blue Mountain Center of Meditation, 2007.
The Holy Bible, New International Version. Grand Rapids: Zondervan House, 1984.
The Holy Qur'an. Translated by Sheikh Muhammad Sarwar. Elmhurst: Islamic Seminary, 1981.
The Myers and Briggs Foundation, 2015, www.myersbriggs.org
Tolkien, J. R. R. *The Lord of the Rings: 50th Anniversary, One Vol. Edition.* New York: Houghton Mifflin Company, 2004.
Trevino, Linda K. and Katherine A. Nelson. *Managing Business Ethics.* Hoboken: John Wiley & Sons, 2011.
Unger, Roberto M. *Knowledge and Politics.* New York: Free Press, 1975.
Unger, Roberto M. *Politics: A Work in Constructive Social Theory* (3 volumes). Cambridge: Cambridge University Press, 1987.
Updike, John. How to love America and leave it at the same time, *The New Yorker*, August 1972.
Verbeek, Bruno and Morris Christopher. Game theory and ethics. In Edward N. Zalta (ed.), *The Stanford Encyclopedia of Philosophy* (Summer Edition), 2010, http://plato.stanford.edu/archives/sum2010/entries/game-ethics/
von Neumann, John and Oskar Morgenstern. *Theory of Games and Economic Organization* (60th Anniversary Commemorative Edition). Princeton: Princeton University Press, 1944/2004.
Walton, Sam with John Huey. *Made in America.* New York: Doubleday, 1992.
Weaver, Gary and Linda Trevino. The boundary between fact and value, *Business Ethics Quarterly*, 4 (1994): 129–143.
Weaver, Gary R. and Michael E. Brown. Moral foundations at work: New factors to consider in understanding the nature and role of ethics in organizations. In David De Cremer and Ann E. Tenbrunsel (eds.), *Behavioral Business Ethics: Shaping an Emerging Field.* New York: Routledge, 2012.
Weber, Max. *The Protestant Ethic and the Spirit of Capitalism.* New York: Routledge, 2001.
Weber, Max. *The Vocation Lectures.* Translated by Rodney Livingstone. Indianapolis: Hackett Publishing Company, 1918/2004.
Welch, Jack and John A. Byrne. *Straight from the Gut.* New York: Warner Books, 2003.
White, Edmund. The man who understood Balanchine, *The New York Times*, November 1998. Accessed on December 15, 2014, https://www.nytimes.com/books/98/11/08/bookend/bookend.html
Williams, Bernard. The human prejudice, *Lecture*, 2002, https://www.youtube.com/watch?v=szgMiqbR57s

Williams, Bernard. *Philosophy as a Humanistic Discipline.* Princeton, NJ: Princeton University Press, 2008.
Williams, Pharrell. Happy, *Columbia Records,* 2013.
Williams, William Carlos. *Paterson.* New York: New Directions, 1927/1995.
Wilson, D. S. Altruism and organism: Disentangling the themes of multilevel selection theory, *The American Naturalist,* 150 (1997): S122–S134.
Wilson, D. S. *The Neighborhood Project: Using Evolution to Improve My City, One Block at a Time.* New York: Hachette Book Group, 2011.
Wilson, E. O. *Sociobiology: The New Synthesis.* Cambridge: Harvard University Press, 1975/2000.
Wittgenstein, Ludwig. *Philosophical Investigations* (3rd ed.). Translated by G. E. M. Anscombe. London: Pearson, 1973.
Wright, Robert. *The Moral Animal: Why We Are, the Way We Are: The New Science of Evolutionary Psychology.* New York: Random House, 1994.
Wright, Robert. *Non-Zero: The Logic of Human Destiny.* New York: Random House, 2000.
Zimbardo, Philip. *The Lucifer Effect: Understanding How Good People Turn Evil.* New York: Random House, 2008.

Index

Academy of Management, 143
Agreeableness (game), 72–73, 91
airport, games played at, 26–27
alternative game-theoretic stories, different interpretations of, 90–91
altruism and egoism, 5, 6–7, 96, 99, 132, 139, 172, 173
altruism, shame in relation to, 98
Altruist's Dilemma (game), 175
Arikha, Noga, 176
Aristotle, 13, 14, 15, 51, 57, 93–95, 145, 147
   definition of virtue by, 57
Arjuna, 141
Assertion (game), 70
Austen, Jane, 3
Austen, Jane (Phlegmatic temperament), 13
Authenticity, 122–123

Balanchine, George, 1–2, 18
Battle of the Sexes, 2, 82–83, 89
Bentham, Jeremy, 13, 15
Berne, Eric, 92
blame, 128, 139, 169–170
Boehm, Christopher, 21, 34
Bowles, Samuel, 34–35
Bryan, Nicole, 134–135
Buffett, Warren (Phlegmatic temperament), 13
business and industry, connection of to modernity, 148–149
business education, 158

business ethics, xi, 1, 15, 159
   academic-practitioner divide in, 121
   as an ascendant worldview, 143, 149, 160, 161
   characteristics of in games, 152
   compared to other ethics, 153, 162
   connection between Sanguine and Phlegmatic Harmony in, 124
   connection of to Four Temperaments approach, 139
   contradiction or tension within, 156–157
   critical, 119–121, 139
   dark side of, 124
   ethical relations approach to, 126, 164–166
   future of, 149–150, 157–158
   institutional context of, 156
   legal approach to, 125, 166–169
   moral parity of with other ethics, 156–157
   normative-empirical divide in, 121
   as one path to Harmony, 56–57
   as a Phlegmatic role ethics, 142, 155
   possibility of universalizing, 157
   psychological approach to, 125
   as singular or plural, 162
   as a subject for young people, 158
   success of different character types within, 134
   teaching of, 125–138
   traditional ethics and, 152
   virtue ethics approaches to, 177

Index

cafes, Harmony games played in, 25–26, 78–79, 153, 156
call centers, 122–124
Camerer, Colin, 182
Camus, Albert, 115
capitalization of games and temperaments, 175
Carver, John, 82
Chicken, 2, 88
Choler, 88, 89, 101–102
　and Disharmony Games, 88
　and Partial Disharmony Games, 89
　and Unequal Imperfect Harmony Games, 101–102
Choleric ethics, 149
Choleric games, 16, 64–65, 91, 103–104, 114
Choleric Harmony, 28, 40–41, 44, 46
Choleric imbalance in Nietzsche and Marx, 147, 148
Choleric reason, 76
Choleric temperament, xix, 4, 13
Cialdini, Robert, 131
Clementi, Tyler, 41
Collins, Suzanne, 43–45, 106, 120
competitive games, 3
competitiveness, 98
competitive striving to be ethically better, 55
Complacency (game), 78–79
Cole, Teju, 27–29, 38
Conan Doyle, Arthur, 18, 111–113
Confucius, 15
consequences, intended and unintended, 139, 169–170
Cook, Tim, 38
credible commitments, 134

Darwin, Charles, 34–35, 94
Dawkins, Richard, 34, 36, 94, 99
Deference (game), 5–6, 7, 14, 62, 65, 67–69, 83, 88, 91, 97
democracy, 145
Dewey, John, 151
dictator games, 30
Disharmony, 7, 45, 105
　different types of, 98

Disharmony Games, 4, 16, 22, 87, 91, 103, 114, 130, 150, 172
　Choler and, 88
　definition of, 163
　HJV and Dominance in opposition in, 64–65
　role of subselves in solving, 103
Dishonesty (game), 72
Dominance (game theory concept), 62–63, 100, 131–132
dominant strategy, 43
Domino's Pizza, 125–126, 164–169
Donaldson, Thomas, 121
Donne, John, 11–12, 18
Dostoevsky, Fyodor, 57
Dunlap, Al, 18

Ebbers, Bernie, 136
economics and psychology, relation between, 122
egoism, competitiveness in relation to, 98
Engels, Friedrich, 147–148
Equal Imperfect Harmony Games, role of melancholy subself in solving, 105
ethical character of nonhuman entities, 111, 114
ethical focal points, 23–25, 128–130, 133, 170–172
ethical relations approach to business ethics, 125–126, 164–166
ethical wisdom of crowds, 132–133, 173
ethics, 8–9
　imperative mood in, 12, 176
　as a successful strategy, 8–9, 176
　as a universal ranking device, 110
evolutionarily stable strategy, 96, 176
exercises, 17–18, 38, 60, 92, 115, 140, 160
Exploitation (game), 74–75, 104

fairness judgments and self-interest, 130–131, 172
fast and frugal approach, 176
fiction and Harmony Games, 43–47, 106, 179

Fish, Stanley, 151
Fiske, Alan, 30
flipped stories, 62, 90–92
Followership (game), 101
Follow the Rule (game), 64, 86, 91
Four Temperaments, xix, 4, 14–15, 60
   Balanchine ballet, 1
   and game-theoretic stories, 90
   historical imbalances in, 149
   ideal of balance among, 12, 161
   and Nature, 114
   relation of to social games,
     16–17, 161
   value of in social games, 9, 99,
     114, 143
Four Temperaments approach, 11, 150
   contrast of to universal ethics,
     150–151
   optimism in, 11
   relation of to business ethics, 139
Frank, Robert, xii, 7, 139
Freud, Sigmund, 10
Friedman, Milton, 18
Fukuyama, Francis, 145–146

game, meaning of, 3–4, 107
game-theoretic stories and the Four
   Temperaments, 90
game theory, 2
   classical, 111
   critical and mainstream, 61, 66
   critical approach to, 5, 66, 109
   evolutionary, 96–97
   flipped, 66
   interpretation of key terms in, 107
   mainstream, 66
games, 3–4
   for classroom use, 127–134
   played in airport, 26–27
   played by bacteria, 105
   played in cafes, 25–26, 78,
     153, 156
   played by mitochondria and
     nuclei, 98
   played by yeast cells, 101, 103, 106
Gandhi, Mahatma, 24
Genesis (Adam and Eve), 177

Gigerenzer, Gerd, 176
Gilligan, Carol, 29
Gintis, Herbert, 34–35
Glaucon (ring story), 57–58
Goodall, Jane, 33–34, 38
governance, Phlegmatic Harmony
   and, 154
Gratitude (game), 78–79, 84, 91
Greene, Joshua, 38, 141–142, 145,
   160, 182

Haidt, Jonathan, xiii, 10, 21, 30, 38,
   142, 146, 160, 180, 182
Hampshire, Stuart, 177
harassment, intimidation, and
   bullying, 41
Harmony, 1
   and ambiguous payoffs, 138
   business ethics as one path to, 56
   construing life in terms of, 155
   dependence of on ability to solve
     difficult games, 33–34, 71
   eightfold division of, 39–43, 59
   as good, case for, 50–52
   and manipulation, 135–138
   and moral wrong, 39, 49–50, 138
   and writing, 51
Harmony Games, 4, 13–14, 16, 21, 22,
   25, 28, 37, 39, 91, 99, 105, 114,
   140, 150, 161
   creating, 48, 135–138
   definition of, 163
   and fiction, 43–47, 106, 179
   HJV and Dominance in accord
     in, 63
   how different from win-win games,
     48–49
   inevitability of, 123
   Sanguine enjoyment and, 85
   types of, 39–43, 59
   writing as a universal form of, 56
Harmony perspective on human
   nature, 53–54
Hegel, G.W.F., 144–146
Henrich, Joseph, 30, 38, 182
heuristics and biases approach, 176
high culture, 120

Index

Highest Joint Value (HJV), 27, 62–63, 76–77, 99–102, 130–133
   Benthamite interpretation of, 108
   individualist interpretation of, 108–109
   Rawlsian interpretation of, 109
history of ethics, 159
history, idealist and materialist accounts of, 145, 147–149
Hobbes, Thomas, 30–31
Holmes, Sherlock, 111–113
Holzinger, Katherina, 179
Holocaust, 29
homo economicus, 97
human moral psychology as a vast conspiracy to make us Harmonize, 21, 129–130
human nature, negative and positive components of, xix, 10–11, 16
human prehistory, 32–33, 36, 178
Human Relations management school, 67
Hume, David, 8, 10, 18, 52, 119
   distinction between "is" and "ought" by, 52–53
humors, 8–9
Huntington, Samuel, 146
hypersociality, 35

id-ego-superego, contrasted to Four Temperaments, 10–11
Idiots! (game), 76–77
I Get it! (game), 6, 76–77, 91
ignorance is bliss, 22
Ignorance Is Bliss (figure), 37
Imperfect Harmony Games, 16, 22, 64, 86, 91, 100, 104, 114, 150
   definition of, 163–164
   Phlegmatic temperament and, 86, 89
Invisible Hand (game), 85
invisible hand story, 78

James, Henry, 45–47, 106, 120
James, William, 151
Jensen, Michael, 72–73
Jobs, Steve, 18

Kahneman, Daniel, 10, 176
Kali (Choleric Temperament), 13
Kant, Immanuel, 15, 31, 176, 178
Kelleher, Herb, 48–49
Kelman, Mark, 176
Kennedy, Duncan, xii, 119, 151
King, Martin Luther, 24
Knobe effect, 62–63, 128, 139, 169–170
Knobe, Joshua, 128, 139
Kohlberg, Lawrence, 29
Krishna, 141
Kurzban, Robert, 181

Lagos, 27–28
Lao Zi, 15, 65
leadership, 83
   and followership, 154, 173–174
   and the Phlegmatic style, 134
   Sanguine, 125, 133
   types of, 133
Leadership (game), 83, 101
Leadership/Followership (game), 101
Legal Realist movement, 151
Lincoln, Abraham, 13, 24, 25, 129
Locked-in (game), 103
Love (game), 79–80, 84, 91, 106

MacIntyre, Alasdair, 94–95
Management, economic and psychological approaches to, 122–123
Managerialism (game), 6, 65, 80–81, 84, 91, 104, 154–155
managers, use of System 1 and System 2 by, 80
Manson, Charles, 24, 129
Mary (Melancholy Temperament), 13
Marx, Karl, 144–148
master-slave dialectic (Hegel), 145–146
Melancholy ethics, 149–152
Melancholy games, 16, 64, 91, 104, 114
Melancholy Harmony, 41, 44
Melancholy temperament, xix, 4–5, 13
Mellowness (game), 85, 91
Milgram, Stanley, 29, 136
MineCraft, 3
Monaghan, Tom, 126, 164–166

Moral Foundations Theory (Haidt), 30
motivations, social and egoistic, 5
Myxococcus bacterium, 105

Nash, John Forbes, 2, 18
natural selection, logic of, 95–96, 112–113
Nature, 99
Nature and the Four Temperaments, 114
negativity bias, 50
negative and positive emotions, 50
Newark, 28
Nietzsche, Friedrich, 13, 57, 144–147
1970s pessimism on human nature, 29
Nozick, Robert, 179

obedience-oriented ethics, 150

Pacifist's Dilemma (game), 32, 66
Partial Disharmony Games, 16, 64, 88, 91, 104, 114
  Choler and, 89
  definition of, 163
Parts and wholes, logic of, 98
Pascal, Blaise, 13, 65–66
payoffs, 107
  ambiguous or unknown, 22–23, 25, 37, 54–55, 102
  desirability of ambiguity in, 22, 37
  known, 22–23, 37, 102
  meaning of, 107
personality psychology, 12–13
Phlegmatic games, 16, 64, 91, 100, 104, 114, 160
Phlegmatic Harmony, 42, 44, 46, 120, 153–154, 159
  and governance, 154
Phlegmatic temperament, xix, 4, 13
  and Imperfect Harmony Games, 86, 89, 153–154
  and leadership, 134
Pinker, Steven, 31–33, 146
Plato, 15, 57, 145
player, meaning of in game theory, 107
policy governance, 82

political ethics, 151
political ideology, xiii
popular culture, 120
population growth, 178
premodern medicine, 9
Price, George, 178
priestly ethics, 15, 149–150, 159
  contradiction or tension within, 157
Prisoner's Dilemma, xi, xii, 2, 4, 7, 14, 23, 32, 62, 66, 87, 129–131, 172, 176
  absence of the Sanguine in, 67
pro-social motivations, 62
psychology and economics, relation between, 122
Public Spaces Project, 38
Purposive cosmos, 14

Rapoport, Anatol, 92, 179
Rationality (game), 73–76, 91, 103
Reagan, Ronald (Sanguine temperament), 13
reason and passion, 10–11
reasoning, 50–51, 94
resentment, 144, 146
Respect (game), 79, 106
Retaliation against players who do not play HJV, 104
Roosevelt, Franklin (Sanguine temperament), 13
Rorty, Richard, 151
Rubin, Gretchen, 180
Russell, Bertrand, 57

Sanguine ethics, 15, 159
  possible future rise of, 149–150
Sanguine games, 16, 91, 114
Sanguine Harmony, 42, 44, 47, 124, 130, 157
  moving from Phlegmatic Harmony to, 76, 78–79
Sanguine leadership, 125, 133
Sanguine reason, 59, 62, 70, 76, 90, 92, 94, 113, 121, 128
  assumption of good character in, 71–72

Index

Sanguine System 2 Harmony, difficulty in attaining, 39, 42, 59
Sanguine temperament, xix, 4, 11, 13, 14, 22, 39
Schelling, Thomas, xi, xiv, 2, 18, 23–24, 61, 62, 119, 128, 131, 134, 139, 163–164
Seidman, Dov, 121
self-expression, 124
self-interest and fairness judgments, 130–131, 172
self-sacrifice, 82, 132
Self-Sacrifice (game), 91, 101
selfish gene, 34, 99
Sensitive Boy (game), 64, 89, 91, 101–102
shame, 98
Shweder, Richard, 30
Sim City, 3
Singer, Peter, 36
Skyrms, Brian, 105
slave morality (Nietzsche), 144–145
Smith, Adam, 6, 78, 150, 176
Sober, Elliott, 34
Social Darwinism, 97
Society for Business Ethics, 143
social games, xix, 3, 99
    historical modes of solving, 143–144
social insects, 35
Socrates, 7–10, 51, 57
species-ism, 21, 36, 178
Staats, Bradley, 122
Stag Hunt, 2, 82, 86, 105
Stranger Mother (game), 87, 91, 103
strategy, meaning of in game theory, 107–108
subselves, 100
    acting and reacting, xix, 16, 100, 114
    contradictions or tensions among, 110–111
    negative and positive, xix, 16, 100, 114
Sullivan, Scott, 126, 136–137
Surowiecki, James, 132
System 1 and System 2 (Kahneman), 10, 59, 69, 80, 120
System 1 and System 2 Harmony, 47, 59–60, 127
System 2 elites, 158–159

Taylor, Frederick, 67
technology, 120
telos (purpose), 14, 93, 95, 113–114, 180
The Dhammapada, 177
The Bhagavad Gita, 141, 177
The Koran, 178
The Last Shall Be First (game), 82, 91, 105
The Sermon on the Mount, 177–178
Tit for Tat, 176
Tolkien, J.R.R., 58
Tolstoy, Leo, 115

ultimatum games, 30
Unequal Imperfect Harmony Games, role of Choler in solving, 101–102
Unger, Roberto, 151
universe, undersocialization or oversocialization in, 97
Updike, John, 21, 38

Value Competition (blog), 160, 180
value of being extreme, 69
vengeance, 33–34, 104
    high desire for by humans compared to chimpanzees, 33–34
Vinson, Betty, 126, 136–138
virtue, 57
    Aristotle's definition of, 57
    cultivation of, 70
virtue ethicists, 15

Wal-Mart, 126
warrior ethics, 14–15, 149, 159
    contradiction or tension within, 156–157
Weber, Max, 119
Welch, Jack, 18, 48–49
    Choleric temperament in, 13
why business ethics matters, two perspectives on, 142–143
Williams, Bernard, 178–179
Wilson, David Sloan, 34–35, 95
Wilson, E.O, 21, 34–35

Wipro, 122–124
wisdom of crowds, 132, 173
Wittgenstein, Ludwig, 3
WorldCom, 126, 136–138
World War II, 2, 176
Wright, Robert, 180
writing, 51
    difficulty of Harmonizing in, 51
    experimentation in, 121

rise of, 51
as a universal form of Harmony Game, 56

yin and yang (Lao Zi), xix, 16, 65, 91, 114

Zeus (Sanguine temperament), 13
Zimbardo, Philip, 29

CPSIA information can be obtained
at www.ICGtesting.com
Printed in the USA
LVOW04*1122170917
549029LV00010B/182/P